Textbook of Animal Husbandry

Textbook of Animal Husbandry

– Author –
Chandrama Singh

Wordpen Academics
Publishing Science That Matter

ISBN: 978-93-90438-91-4 (Hardback)

Published by : **BIO-GREEN BOOKS**
4736/23, Ansari Road, Darya Ganj,
New Delhi – 110 002
Phone: 011-23245578, 23244987
E-mail: biogreenbooks@gmail.com

Printed at : Replika Press Pvt. Ltd.

Preface

A large number of farmers in India depend on animal husbandry for their livelihood. In addition to supplying milk, meat, eggs, and hides, animals, mainly bullocks, are the major source of power for both farmers and drayers. Thus, animal husbandry plays an important role in the rural economy. The gross value of output from this sector was Rs 358 billion in FY 1989, an amount that constituted about 25 percent of the total agricultural output of Rs 1.4 trillion. India has 30 per cent of the world's cattle. There are 26 distinctive breeds of cow in India. The hump, long ears and bushy tail distinguish the Indian cow. Here cows are everywhere! Because the cow is respected as a sacred animal, it's allowed to roam unharmed, and they are pretty used to the traffic and the rhythm of the city. So, you can see them roaming the streets in towns and cities, grazing unmindfully on the roadside grass verges and munching away veges thrown out by street sellers. Stray and homeless cows are also supported by temples, especially in southern India. Sheep are most likely descended from the wild mouflon of Europe and Asia. One of the earliest animals to be domesticated for agricultural purposes, sheep are raised for fleece, meat (lamb, hogget or mutton) and milk. A sheep's wool is the most widely used of any animal, and is usually harvested by shearing. Ovine meat is called lamb when from younger animals and mutton when from older ones. Sheep continue to be important for wool and meat today, and are also occasionally raised for pelts, as dairy animals, or as model organisms for science. Sheep husbandry is practised throughout the majority of the inhabited world, and has been fundamental to many civilizations. In the modern era, Australia, New Zealand, the southern and central South American nations, and the British Isles are most closely associated with sheep production.

The Livestock Insurance Scheme, a centrally sponsored scheme, which was implemented on a pilot basis during 2005-06 and 2006-07 of the 10th Five Year Plan and 2007-08 of the 11th Five Year Plan in 100 selected districts. The scheme is being implemented on a regular basis in 300 districts of the country. The Livestock Insurance Scheme has been formulated with the twin objective of providing protection mechanism to the farmers and cattle rearers against any eventual loss of their animals due to death and to demonstrate the benefit of the insurance of livestock to the people and popularize it with the ultimate goal of attaining qualitative improvement in livestock and their products. Under the scheme, the indegenous/ crossbred milch cattle and buffaloes are being insured at maximum of their current market price. The premium of the insurance is subsidized to the tune of 50%. The entire cost of the subsidy is being borne by the Central Government. The benefit of subsidy is being provided to a maximum of 2 animals per beneficiary for a policy of maximum of three years. The scheme is being implemented in all states except Goa through the State Livestock Development Boards of respective states.

This book contains the fundamental and basic information of subject and the selection of contents makes it an appropriate book for the students.

— *Dr. Chandrama Singh*

Contents

1

Cow Management

As the sheep is to Christianity, the cow is to Hinduism. Lord Krishna was a cowherd, and the bull is depicted as the vehicle of Lord Shiva. Today the cow has almost become a symbol of Hinduism.

Cows, Cows Everywhere

India has 30 per cent of the world's cattle. There are 26 distinctive breeds of cow in India. The hump, long ears and bushy tail distinguish the Indian cow.

Here cows are everywhere! Because the cow is respected as a sacred animal, it's allowed to roam unharmed, and they are pretty used to the traffic and the rhythm of the city. So, you can see them roaming the streets in towns and cities, grazing unmindfully on the roadside grass verges and munching away veges thrown out by street sellers. Stray and homeless cows are also supported by temples, especially in southern India.

Conserve the Cow

As opposed to the West, where the cow is widely considered as nothing better than walking hamburgers, in India, the cow is believed to be a symbol of the earth- because it gives so much yet asks nothing in return.

Because of its great economic importance, it makes good sense to protect the cow. It is said Mahatma Gandhibecame a vegetarian because he felt cows were ill-treated. Such is the respect for the cow, notes scholar Jeaneane Fowler in her book on Hinduism, that Indians had offered to take in millions of cows waiting for slaughter in Britain as a result of the crisis in beef production in 1996.

Religious Significance of the Cow

Though the cow is held sacred to the Hindus, it is not exactly worshipped as a deity by all. On the 12th day of the 12th month of the Hindu calendar, a cow ritual is performed in Jodhpur palace, in the western Indian state of Rajasthan.

Bull Temples

Nandi Bull, a vehicle of the gods, is considered the symbol of respect for all male cattle. The Nandi Bull holy site at Madurai and the Shiva temple at Mahabalipuram are the most venerated bovine shrines.

Even non-Hindus are allowed to enter the 16th century Bull Temple in Bangalore. The Vishwanath Temple of Jhansi, believed to have been built in 1002, also has a large statue of the Nandi Bull.

History of the Holy Cow

The cow was venerated as the mother goddess in the early Mediterranean civilizations. The cow became important in India, first in the Vedic period (1500-900 BCE), but only as a symbol of wealth. For the Vedic man cows were 'the "real life" substratum of the goods of life', writes JC Heesterman in The Encyclopedia Of Religion, vol. 5.

Cows as Symbol of Sacrifice

Cows form the core of religious sacrifices, for without ghee or clarified liquid butter, which is produced from cow's milk, no sacrifice can be performed.

In the Mahabharata, we have Bhishma saying: "Cows represent sacrifice. Without them, there can be no sacrifice...Cows are guileless in their behaviour and from them flow sacrifices...and milk and curds and butter. Hence cows are sacred..."

Bhishma also observes that the cow acts as a surrogate mother by providing milk to human beings for the whole life. So the cow is truly the mother of the world.

Cows as Gifts

Of all gifts, the cow is still considered the highest in rural India. The Puranas, ancient Hindu scriptures, have it that nothing is more pious than the gift of cows. "There is no gift that produces more blessed merit." Lord Rama was given a dowry of thousands of cows and bullocks when he married Sita.

Cow-Dung, Ahoy

Cows are also thought to be cleansers and sanctifiers. The cow-dung is an efficacious disinfectant and often used as fuel in lieu of firewood. In the scriptures, we find the sage Vyasa saying that cows are the most efficacious cleansers of all.

No Beef Please

Since the cow is thought to be God's useful gift to mankind, consuming beef or veal is considered sacrilegious for Hindus. Selling beef is banned in many Indian cities, and few Hindus would be ready to even taste cattle meat, for socio-cultural reasons.

Breeds of Cattle and their Selection

Indian Cattle Breeds

Milch breeds:

- Mainly found in Punjab, Haryana, U.P, Delhi, Bihar and M.P.
- Milk yield:
  - Under village condition :1350 kgs
  - Under commercial farms: 2100 kgs.
- Age at first calving-32-36 months
- Calving interval – 15 month.

Gir:

- Mainly found in Gir forest areas of South Kathiawar
- Milk yield:
  * Under village condition : 900 kgs
  * Under commercial farms: 1600 kgs.

Tharparkar:

- Mainly found in Jodhpur, Kutch and Jaisalmer
- Milk yield:
  * Under village condition :1660 kgs
  * Under commercial farms: 2500 kgs.

Red Sindhi;

- Mainly available in Punjab, Haryana, Karnataka, Tamil Nadu, Kerala and Orissa.
- Milk yield:
  * Under village condition :1100 kgs
  * Under commercial farms: 1900 kgs.

Milch and Draught Breeds

Ongole:

- Mainly found in Nellore, Krishna, Godavari and Guntur district of A.P.
- Milk yield –1500 kgs
- Bullocks are powerful for cart work and heavy ploughing.

Hariana:

- Mainly found in Karnal, Hisar and Gurgaon district of Haryana, Delhi and western M.P

- Milk yield –1140-4500 kgs
- Bullocks are powerful for road transport and rapid ploughing.

Kankrej:

- Mainly found in Guajrat
- Milk yield:
    * Under village condition :1300 kgs
    * Under commercial farms : 3600 kgs.
- Age at first calving-36 to 42 months
- Calving interval – 15 to 16 months
- Bullocks are fast, active and strong. Good for plough and cart purpose.

Deoni:

- Mainly found in North western and western parts of A.P.
- Cows are good milk producers and bullocks are good for work.

Draught Breeds

Amritmahal:

- Mainly found in Karnataka.
- Best suitable for ploughing and transport.

Hallikar:

- Mainly found in Tumkur, Hassan and Mysore districts of Karnataka

Khillar

Kangayam:

- Mainly found in Coimbatore, Erode, Namakkal, Karur and Dindigul districts of Tamil Nadu.
- Best suited for ploughing and transport. Withstands hardy conditions.

Exotic Dairy Breeds

Jersey:

- Age at first calving : 26-30 months
- Intercalving – 13-14 months
- Milk yield – 5000-8000 kgs
- Dairy milk yield is found to be 20 lts whereas cross bred jersey, cow gives 8-10 lts per day.
- In India this breed has acclimatized well especially in the hot and humid areas.

Holstein Friesian:

- This breed is from Holland
- Milk yield -7200-9000 kgs
- This is by far the best diary breed among exotic cattle regarding milk yield. On an average it gives 25 litres of milk per day, whereas a cross breed H.F. cow gives 10-15 lts per day.
- It can perform well in coastal and delta areas

Breeds of Buffalo

Murrah:

- Mainly found in Haryana, Delhi and Punjab
- Milk yield – 1560 kgs
- On an average the daily milk yield is found to be 8-10 lts, whereas a cross breed murrah buffalo gives 6-8 lts per day.
- It performs well in coastal and slightly cold climatic areas.

Surti:

- Gujarat
- 1700-2500 kgs.

Jaffarabadi :

- Kathiawar district of Gujarat
- 1800-2700 kgs.

Nagpuri:

- Nagpur, Wardha, Akola, Amravathi and Yeotmal in Maharastra.
- Milk yield – 1030-1500 kgs.

General Selection Procedures for Dairy Breeds

Selecting a calf in calf show, a cow in cattle show by judging is an art. A dairy farmer should build up his own herd by breeding his own herd.

Following guidelines will be useful for selection of a diary cow:

- whenever an animal is purchased from a cattle fair, it should be selected based upon its breed characters and milk producing ability
- History sheet or pedigree sheet which are generally maintained in organized farms reveals the complete history of animal
- The maximum yield by dairy cows are noticed during the first five lactations. So generally selection should be carried out during First or Second lactation and that too are month after calving.

- There successive complete milkings has to be done and an average of it will give a fair idea regarding production by a particular animal
- A cow should allow anybody to milk, and should be docile.
- It is better to purchase the animals during the months of October and November.
- Maximum yield is noticed till 90 days after calving.

Breed characteristics of high yielding dairy cows:

- Attractive individuality with feminity, vigour, harmonious blending of all parts, impressive style and carriage
- Animal should have wedge shaped appearance of the body
- It should have bright eyes with lean neck
- The udder should be well attached to the abdomen
- The skin of the udder should have a good network of blood vessels
- All four quarters of the udder should be well demarcated with well placed teats.

Selecting breeds for Commercial Dairy Farm-Suggestions:

- Under Indian condition a commercial dairy farm should consist of minimum 20 animals (10 cows, 10 buffaloes) this strength can easily go up to 100 animals in proportion of 50:50 or 40:60. After this however, you need to review your strength and market potential before you chose to go for expansion.
- Middle class health-conscious Indian families prefer low fat milk for consumption as liquid milk. It is always better to go for a commercial farm of mixed type. (Cross breed, cows and buffaloes kept in separate rows under one shed).
- Conduct a through study of the immediate market where you are planning to market your milk You can mix milk from both type of animals and sold as per need of the market. Hotels and some general customers (can be around 30%) prefer pure buffalo milk. Hospitals, sanitariums prefer cow's milk.

Selection of Cow/Buffalo Breeds for Commercial Farm

Cows:

- Good quality cows are available in the market and it cost around Rs.1200 to Rs.1500 per liter of milk production per day. (e.g. Cost of a cow producing 10 liter of Milk per day will be between Rs.12,000 to Rs.15,000).
- If proper care is given, cows breed regularly giving one calf every 13-14 month interval.
- They are more docile and can be handled easily. Good milk yielding cross breeds (Holstein and Jersey crosses) has well adapted to Indian climate.

- The fat percentage of cow's milk varies from 3-5.5% and is lower then Buffaloes.

Buffaloes:

- In India, we have good buffalo breeds like Murrah and Mehsana, which are suitable for commercial dairy farm.
- Buffalo milk has more demand for making butter and butter oil (Ghee), as fat percentage in milk is higher then cow. Buffalo milk is also preferred for making tea, a welcoming drink in common Indian household.
- Buffaloes can be maintained on more fibrous crop residues, hence scope for reducing feed cost.
- Buffaloes largely mature late and give birth to calves at 16 to 18 months interval. Male calves fetch little value.
- Buffaloes need cooling facility e.g. wallowing tank or showers/foggers with fan.

Foot and Mouth Disease

Foot and mouth disease (FMD) is a highly contagious disease of cloven footed animals including cattle, sheep, goat and pigs. This disease is endemic in India and producing serious economic losses to the country due to the ban on export of livestock products to other countries and reduced productivity of diseased animals.

What are the symptoms?

- High fever
- Decreased milk production
- Appearance of vesicles on the foot, mouth and in udder
- lameness due to the presence of vesicles in the foot
- Excessive frothy salivation due to the presence of vesicles in the mouth.

Disease Symptoms in the Mouth

Disease Symptoms in the Foot

How this disease spread?

- The virus is excreted in the secretions and excretions of the affected animals including saliva, milk and vesicular fluid
- The virus spread from one place to another by air especially when humidity of the air is high the virus easily spread from one place to another.
- The disease also spread from affected animals to healthy animals by contaminated feed, water and other farm appliances and also through the movement of animals such as dogs, birds and farm workers

- Infected sheep and pigs are excreting higher amount of infective virus and they play an important role in the spread of this disease
- Cross bred cattle are highly susceptible than the indigenous cattle
- The disease also spread by transportation of affected animals from one place to another.

What are the after-effects?

Diseased animals are more susceptible to mastitis, failure of conception, reduced heat tolerance and decreased milk production

How to control the spread?

- The healthy susceptible animals should not be transported to the endemic areas
- Animals should not be purchased from disease affected areas
- Newly purchased animals should be quarantined for 21 days from the other animals in the farm Treatment
- Mouth and foot of the diseased animals washed with 1 % potassium permanganate solution. Antiseptic lotion can be applied to the vesicles in the foot. Boric acid glycerine paste can be applied to the vesicles in the mouth.
- Diseased animals should be fed with palliative diet and they should be isolated from the healthy animals.

Vaccination Schedule:

- Vaccination of all susceptible animals with FMD vaccine once in 6 months. Vaccination programme should include cattle, sheep, goat and pigs.
- First vaccination to the calves should be given at the age of 4 months and second vaccine should be given at 5th month. Thereafter booster doses should be given once in 4-6 months.

2

Buffalo Farm Management

The buffalo is a multipurpose animal. Not only is it a better source of milk than the cow, it also provides meat and works as a natural source of energy.

Water Buffalo

The water buffalo or domestic Asian water buffalo (Bubalus bubalis) is a large bovine animal, frequently used as livestock in southern Asia, and also widely in South America, southern Europe, north Africa, and elsewhere.

In 2000, the United Nations Food and Agriculture Organization estimated that there were approximately 158 million water buffalo in the world and that 97% of them (approximately 153 million animals) were in Asia.

There are established feral populations in northern Australia but the dwindling true wild populations are thought to survive in India, Bangladesh, Pakistan, Nepal, Bhutan, and Thailand. All the domestic varieties and breeds descend from one common ancestor, the Wild Water Buffalo, which is now an endangered species. The domestic Water Buffalo, although derived from the Wild Water Buffalo, is the product of thousands of years of selective breeding in either South Asia or Southeast Asia.

Buffalo are used as draft, meat, and dairy animals. Their dung is used as a fertilizer and as a fuel when dried. In Chonburi, Thailand, Pakistan and in South western region of Karnataka, India, there are annual water buffalo races known as Kambala. A few have also found use as pack animals carrying loads even for special forces.

American bison are known as buffalo in parts of North America, but not normally in other usages; bison are more closely related to cattle, gaur, banteng, and yaks than to Asian buffalo. The water buffalo genus includes water buffalo, tamaraw and anoas—all Asian species. The ancestry of the African buffalo is unclear, but it is not believed to be closely related to the water buffalo.

Wild Water Buffalo

True wild water buffalo are thought to survive in Bangladesh, India, Nepal, Bhutan, and Thailand.

The IUCN Red List of threatened species classifies wild water buffalo (Bubalis arnee) as an endangered species in Southeast Asia. They run rampant as a wild animal in Western Australia, which is the only part of the world where they are legally hunted, and because of their large numbers there, encouraged to do so by the government, in their original range. The total number of wild water buffalo left in Southeast Asia, is thought to be less than 4,000, which suggests that the number of mature individuals will be less than 2,500, and an estimated continuing decline of at least 20% within 14 years (ca. 2 generations) and at least 50% within 21 years seems likely given the severity of the threats, especially hybridization with the abundant domestic Asian water buffalo leading to genetic pollution.

Anatomy and Morphology

Adult Water Buffalo range in size from 400 to 900 kg (880 to 2,000 lb) for the domestic breeds, while the wild animals are nearly 3 m (9.8 ft) long and 2 m (6.6 ft) tall, weighing up to 1,200 kg (2,600 lb); females are about two-thirds this size.

River buffalo are usually black and have long curled horns, whereas swamp buffalo can be black or white, or both, with gently curved horns. The largest recorded horns are just under 2 metres long.

There are differences between swamp buffalo and river buffalo. Swamp buffalo have swept back horns and are native to the eastern half of Asia from India to Taiwan. All are similar in general appearance. River buffalo generally have curved horns and are native to the western half of Asia.

The rumen (the first chamber of the digestive system of a ruminant) of the Water Buffalo has important differences from that of other ruminants. It consists of essential microorganisms; namely bacteria, protozoa and fungi which digest the food to produce fermentation end-products via anaerobic fermentation or Embden-Myerhof pathway.

The Water Buffalo rumen has been found to contain a larger population of bacteria particularly the cellulolytic bacteria, lower protozoa and higher fungi zoospores. In addition, higher rumen ammonia nitrogen (NH_3-N) and higher pH have been found as compared to those in cattle.

Taxonomy

The classification of the water buffalo is uncertain. Some authorities list a single species, Bubalus bubalis with three subspecies, the river buffalo (B. bubalis bubalis) of South Asia, the carabao or swamp buffalo (B. bubalis carabanesis) of the Philippines and Southeast Asia, and the arni, or wild water buffalo (B. bubalis arnee). Others regard these as closely-related but separate species. The swamp buffalo is primarily found in the eastern half of Asia and has 48 chromosomes. The river buffalo is mostly found in the western half of Asia (and in Europe and Africa), and has 50 chromosomes. The two types do not readily interbreed, but fertile offspring can occur. Buffalo-cattle hybrids have not been observed to occur, and the embryos of such hybrids do not reach maturity in laboratory experiments.

Evolution

Geologically speaking, the Bovidae is much recent group as compared to Cervidae because their members are untraceable in the layers of the earth. The fossil forms of the buffalo provide a definite link between the Indian type and their present extreme representatives and their extinct allies. All Asiatic buffaloes seem to form a closely allied group of species which represent more or less a passage from one variety to another.

Distribution

Type Locality: "Habitat in Asia, cultus in Italia". Restricted by Thomas (1911a:154) to Italy, Rome, but Linnaeus' (1758) comment indicates Asia (India?).

Distribution: Bangladesh, Burma, Cambodia, India (survives in Assam and Orissa), Nepal, N Thailand, Vietnam, and possibly at least formerly in Laos; domesticated in N Africa, S Europe, and even England, east to Indonesia and in E South America; supposedly feral populations in Sri Lanka, Sumatra, Java, Borneo, Philippines and other parts of SE Asia; feral populations resulting from introductions in New Britain and New Ireland (Bismarck Arch., Papua New Guinea), and Australia. Status: CITES – Appendix III (Nepal) as B. arnee (excludes domesticated forms-but see comments below; IUCN – Endangered

Average Lifespan in Captivity: Up to 25 Years

Asia

Asia is the native home of the water buffalo, with 95% of the world population of water buffalo, with about half of the total in India. Many Asian countries depend on the water buffalo as its primary bovine species. It is valuable for its meat and milk as well as the labour it performs. As of 1992 the Asian population was estimated at 141 million.

The fat content of buffalo milk is the highest amongst farm animals and the butterfat is a major source of ghee in some Asian countries. Its success in Asia is evident by its extensive range. Both variants occur in Asia. River buffalo are found in elevations of 2,800 m in Nepal, and swamp buffalo are found throughout the lowland tropics. Part of their success is due to their ability to thrive on poor foodstuffs and yet be valuable economically. Moreover they are much better suited to plough the muddy paddy fields as they are better adapted than common cattle (Bos taurus) to move in swamps. In India the meat of buffalo is sold as beef because it is regarded as different from a cow.

Australia

Swamp buffalo were introduced into the Northern Territory from Timor early in the 19th century as a food source and a beast of burden. They escaped, thrived and became feral, causing significant environmental damage. Buffalo are also found in Arnhem Land and the Top End. An estimated 350,000 buffalo were living on the floodplains of Arnhem Land and the Katherine region in the 1980s. As a result of this they were hunted in the Top End from 1885 until 1980. The commencement of the Brucellosis and Tuberculosis Campaign (BTEC) saw a huge culling program reduce buffalo herds to a fraction of the numbers that were reached in the 1980s. The BTEC was finished when the Northern Territory was declared free of the disease in 1997.

During the 1950s buffalo were hunted for their skins, and meat which was exported and used in the local trade. In the late 1970s live exports were made to Cuba and continued later into other countries. Buffalo are now crossed with riverine buffalo in artificial breeding (AI) programs and may be found in many areas of Australia. Some of these crossbreds are used for milk production.

Melville Island is a popular hunting location, where a steady population of up to 4,000 individuals exist. Safari outfits run out of Darwin to Melville Island and other locations in the Top End often with the use of bush pilots. The horns which can measure up to a record of 3.1 metres tip to tip are a prized hunting trophy.

The buffalo have developed a different appearance from the Indonesian buffalo from which they descend. They live mainly in freshwater marshes and billabongs, and their territory range can be quite expansive during the wet season. Their only natural predator in Australia are large adult saltwater crocodiles, with whom they share the billabongs.

Europe and Middle East

Introduced into North Africa and the Near East by 600 AD, the water buffalo was brought to Europe with returning Crusaders in the Middle Ages, and herds can be found in Turkey, Bulgaria, Romania, Albania, Hungary, Austria and Italy. As in Asia, buffalo of the Middle East and Europe live on coarse vegetation on the marginal land traditionally available to peasants. They are an economic asset by serving as a protein source, draft animal, and storage of family or household wealth. In some areas, they also provide occasional recreation at annual racing festivals. These buffalo are mostly river buffalo; due to genetic isolation, they have adopted a distinct appearance. Buffalo milk is used for the production of buffalo mozzarella in Campania and many other locations around the world.

Egypt

Water buffalo are a traditional farm animal in Egypt, which has a large number of them. They are used as the main source of red meat in Egypt. Cows have been introduced in modern farms, yet water buffalo remain as the more widespread type of cattle in Egypt.

North America

There are very limited commercial herds in North America, for yogurt and cheese products.

South America

Water buffalo were introduced into the Amazon River basin in 1895. They are now extensively used there for beef and dairy production. In 2005, the buffalo herd in the Brazilian Amazon stood at approximately 1.5 million head, of which approximately 160,000 were located in the Lower Amazon floodplain.

Importance to Humans

Water buffalo have been domesticated for 5,000 years and have become economically important animals. They provide more than 5% of the world's milk supply and 20% to 30% of the farm power in Southeast Asia. Milk from these animals is used by many human populations, and is the traditional raw material for mozzarella cheese and curd due to its higher fat content. In Africa and other locations, water buffalo milk is used for yogurt, as in Vermont, USA. The chief dairy breed of Buffalo is the Murrah breed.

Buffalo meat, sometimes called "Carabeef", is often passed off as beef in certain regions and is also a major source of export revenue for India which has the largest population of buffalo in the world. However, in many Asian regions, buffalo meat is less preferred due to its toughness, however, recipes have evolved (Rendang for example) where the slow cooking process and spices not only make the meat palatable, but also preserves it; an important factor in hot climates where refrigeration is not always available.

Water buffalo horns are used for the embouchure of musical instruments such as ney and kaval. Water buffalo hide provides a tough and useful leather often used for shoes and motorcycle helmets. The bones and horns are often made into jewellery, especially earrings.

The water buffalo has promise as a major source of meat, even the milking ones. The water buffalo also is the classic work animal in Asia, an integral part of that continent's traditional village farming structure and also used for hauling cotton, pumping water in Pakistan and hauling logs in Turkey. The domesticated water buffalo is often referred to as "the living tractor of the East" as it is relied upon for plowing and transportation in many parts of Asia.

Nutrition

Milk Composition Analysis, per 100 grams

Constituents	*unit*	*Cow*	*Goat*	*Sheep*	*Buffalo*
Water	g	87.8	88.9	83.0	81.1
Protein	g	3.2	3.1	5.4	4.5
Fat	g	3.9	3.5	6.0	8.0
Carbohydrate	g	4.8	4.4	5.1	4.9
Energy	kcal	66	60	95	110
	kJ	275	253	396	463
Sugars (Lactose)	g	4.8	4.4	5.1	4.9
Fatty Acids:					
Saturated	g	2.4	2.3	3.8	4.2
Mono-unsaturated	g	1.1	0.8	1.5	1.7
Polyunsaturated	g	0.1	0.1	0.3	0.2
Cholesterol	mg	14	10	11	8
Calcium	iu	120	100	170	195

Adaptation and Behavioural

Water buffalo spend much of their day submerged in the muddy waters of Asia's tropical and subtropical forests. They have wide-splayed hoofed feet which are used to prevent them from sinking too deeply in the mud. These adaptations allow them to move in wetlands and swamps. Water buffalo also prefer to feed in grass lands on grass and herbs. Water buffalo behaviour sometimes differs from cattle. For example, most water buffaloes are not trained to be driven. Instead, the herdsman must walk alongside or ahead of them. They then instinctively follow. They also rub against trees more often than cattle do, and they sometimes de-bark the trees causing them to die.

Reproduction

The water buffalo has a reputation for being a sluggish breeder. Without reasonable nutrition, the animals cannot reach puberty as early in life as genetic capability would normally allow. Females normally produce calves every other year after gestation of 9 to 11 months. Young bulls typically remain with maternal herds which consist of around 30 buffalo, for three years after birth. They then go on to form small all-male herds.

Environmental Effects

The water buffalo may affect the environment in either positive or negative ways. Wildlife and conservation scientists have started to recommend and use introduced populations of feral water buffalo to manage uncontrolled vegetation growth in and around natural wetlands. Introduced Asian Water Buffalo at home in such environs provide cheap service by regularly grazing uncontrolled vegetation and opening up clogged water bodies for waterfowl, wetland birds and other wildlife. Grazing water buffalo are sometimes used in Great Britain for conservation grazing, for example to manage Chippenham Fen NNR. These buffalo have been found to be better suited to the wet conditions and poor-quality vegetation than many cattle.

However, in uncontrolled circumstances, water buffalo can cause environmental damage, such as trampling vegetation, disturbing bird and reptile nesting sites, and spreading exotic weeds.

Research

First cloned buffalo: On 15 September 2007, the Philippines announced its development of Southeast Asia's first cloned buffalo. The Philippine Council for Agriculture, Forestry and Natural Resources Research and Development (PCARRD), under the Department of Science and Technology in Los Baños, Laguna approved this project. The Department of Agriculture's Philippine Carabao Center (PCC) will implement "Cloning through somatic cell nuclear transfer as a tool for genetic improvement in water buffaloes". "Super buffalo calves" will be produced. There will be no modification or alteration of the genetic materials, as in GMOs (genetically modified organisms).

Super Carabao

On 1 January 2008, the Philippine Carabao Center in Nueva Ecija, per Filipino scientists, initiated a study to breed the super water buffalo that could produce 4 to 18 litres of milk/day (gene-based technology). Also, the first test-tube hybrid carabao was born there in 2004, named "Glory" after President Gloria Macapagal-Arroyo. Joseph Estrada's most successful project as an opposition senator, the PCC was created through Republic Act 3707, the Carabao Act of 1992.

Some Popular Water Buffalo Festivals

Fighting festivals:

- *"Moh juj" Water Buffalo fighting in Bhogali Bihu of Assam* Moh juj *is held every year in Bhogali Bihu in Assam. Ahotguri in Nagaon is famous for it.*
- "Do Son" Water Buffalo Fighting Festival of Vietnam is held each year on the 9th day of the 8th month of the lunar calendar at Do Son Township, Haiphong City in Vietnam. It is one of the most popular Vietnam festivals & events in Haiphong City. The preparations for this buffalo fighting festival begin right from the 5th and the 6th lunar month itself. The competing buffalo are selected and methodically trained months in advance. It is a

traditional festival of Vietnam attached to a Water God worshipping ceremony and the "Hien Sinh" custom to show marital spirit of the local people of Do Son, Haiphong.

- "Hai Luu" Water Buffalo Fighting Festival of Vietnam According to ancient records, the buffalo fighting in Hai Luu Commune has existed from the 2nd century B.C. General Lu Gia at that time, had the buffalo slaughtered to give a feast to the local people and the warriors, and organized buffalo fighting for amusement. Eventually, all the fighting buffalo will be slaughtered as tributes to the deities.
- "Ko Samui" Water Buffalo Fighting Festival of Thailand is a very popular event held on special occasions such as New Year's Day in January, and Songkran in mid-April, this festival features head-wrestling bouts in which two male Asian water buffalo are pitted against one another. Unlike in Spanish Bullfighting, wherein bulls get killed while fighting sword-wielding men, Buffalo Fighting Festival held at Ko Samui, Thailand is fairly harmless contest. The fighting season varies according to ancient customs & ceremonies. The first Buffalo to turn and run away is considered the loser, the winning buffalo becomes worth several million baht. Ko Samui is an island in the Gulf of Thailand in the South China Sea, it is 700 km from Bangkok and is connected to it by regular flights.

Racing Festivals:

- Chon Buri Water buffalo racing festival, Thailand: Thousands of people flock to this entertainment in downtown Chonburi, 70 km (44 miles) south of Bangkok, at the annual water buffalo festival. About 300 buffalo race in groups of five or six, spurred on by bareback jockeys wielding wooden sticks, as hundreds of spectators cheer. The water buffalo has always played an important role in agriculture in Thailand. For farmers of Chon Buri Province, near Bangkok, it is an important annual festival, beginning in mid-October. It is also a celebration among rice farmers before the rice harvest. At dawn, farmers walk their buffalo through surrounding rice fields, splashing them with water to keep them cool before leading them to the race field. This amazing festival started over a hundred years ago when two men arguing about whose buffalo was the fastest ended up having a race between them. That's how it became a tradition and gradually a social event for farmers who gathered from around the country in Chonburi to trade their goods. The festival also helps a great deal in preserving the number of buffalo, which have been dwindling at quite an alarming rate in other regions. Modern machinery is rapidly replacing buffalo in Thai agriculture. With most of the farm work mechanized, the buffalo-racing tradition has continued. Racing buffalo are now raised just to race; they don't work at all. The few farm buffalo which still do work are much bigger than the racers because of the strenuous work they perform. Farm buffalo are in the "Buffalo Beauty Pageant", a Miss Farmer beauty contest and a comic buffalo costume contest etc.. This festival perfectly exemplifies a favoured Thai attitude to life — "sanuk," meaning fun.

- Babulang Water buffalo racing festival, Sarawak, Malaysia: Babulang is the largest or grandest of the many rituals, ceremonies and festivals of the traditional Bisaya (Borneo) community of Limbang, Sarawak. Highlights are the Ratu Babulang competition and the Water buffalo races which can only be found in this town in Sarawak, Malaysia.
- Vihear Suor village Water buffalo racing festival, Cambodia: Each year, millions of Cambodians visit Buddhist temples across the country to honour their deceased loved ones during a 15-day period commonly known as the Festival of the Dead but in Vihear Suor village, about 22 miles northeast of the capital of Cambodia, Phnom Penh, citizens each year wrap up the festival with a water buffalo race to entertain visitors and honour a pledge made hundreds of years ago. There was a time when many village cattle which provide rural Cambodians with muscle power to plough their fields and transport agricultural products died from an unknown disease. The villagers prayed to a spirit to help save their animals from the disease and promised to show their gratitude by holding a buffalo race each year on the last day of "P'chum Ben" festival as it is known in Cambodian. The race draws hundreds of spectators who come to see riders and their animals charge down the racing field, the racers bouncing up and down on the backs of their buffalo, whose horns were draped with colorful cloth.
- Karapan Sapi racing festival, Madura, Indonesia: The Maduranese people of the island of Madura, East Java, Indonesia, race their strongest and fastest buffalo in races hold regularly a few times a year, typically in August, September and October. It is a very popular spectacle in the towns of Pamekasan, Sampang, and Bangkalan. Besides the prize (and the pride that comes with it), buffalo that win a race are regarded very valuable and are a lot more expensive than their peers. This motivates the owners to feed their buffalo unusual cocktails of high calorie food comprising raw eggs, honey, and herbs in addition to their regular training regime to give them the edge.
- Kambala races, Karnataka, India: The Kambala water buffalo races of Karnataka, India take place between December and March. The races are conducted by having the water buffaloes run in long parallel slushy ditches where they are driven by men standing on wooden planks drawn by the buffaloes. The objective of the race being to finish first and to raise the water to the greatest height.

Buffalo-Many Aspects, One Animal

Of all the domestic animals, the Asian water buffalo holds the greatest promise and potential for production. The buffalo had been severely neglected by the authorities until 1974, when the Food and Agriculture Organization (FAO) first signposted the water buffalo as the most neglected animal. In the intervening years, FAO has more than made up for it by providing leadership and inspiration to national and international organizations in developing water buffalo production.

Still, much knowledge and original thought remains locked in the secretive silence of the non-communicator. What Peter Curtis calls the "failure-to-publish" syndrome, impeding the transfer of technology.

Milk, buffalo: The production of buffalo milk in the Asian-Pacific region exceeds 45 million tonnes annually of which over 30 million tonnes are produced in India alone. With selective breeding, improved management and the establishment of more dairy herds, milk yields are increasing worldwide. The individual 3,000 litre-per-lactation female, considered a record 30 years ago, is now common. There are many which yield 4,000 litres in a lactation of 300 days. And, some have even attained a record of 5,000 litres. The potential for increased milk production, therefore, exists.

The high-yielding Murrah is the Holstein-Friesian of the buffalo world. There is no reason why the Murrah and other riverine breeds should not exceed the volumetric yield of the best milk cows. Buffalo milk contains about twice as much butterfat as cow milk. There is no physiological need for concentrate feed to maintain this level. When concentrates are fed, the fat content rises. When fed ad lib, it can exceed 15 per cent. Fed to excess, this figure can be much higher. The buffalo, in fact, releases unwanted fat in the milk and stores only a minimum in the body tissues.

Work, buffalo: Buffaloes are labour intensive and cost-effective. They are the most versatile of all work animals in the variety of tasks which they can be taught to undertake. All buffalo breeds have a strong milk/meat entity. However, presently the output of thousands of buffaloes is in the form of work energy rather than the direct provision of food as milk or meat. In economic terms, buffaloes are productive and efficient, especially in those agricultural countries where there is ample manpower and little motivation for mechanization.

Meat, buffalo: Recent investigations and trials have convincingly established the fact that when a buffalo is fed well and managed for early slaughter (at a liveweight of 350 to 450 kg), a yield of palatable, high-grade meat can be obtained at a competitive cost. A buffalo carcass has less fat, less bone and a higher proportion of muscle than a cattle carcass. Furthermore a buffalo carcass does not need the customary and costly chill hanging for a specific period to ensure the "eatability" of the product.

It is well known that the buffalo is remarkable for its feed conversion ability, but we do not yet understand how, or why, or whether that capacity can be further improved. The young buffalo calf achieves a daily weight gain of 800 grams without any supplement feed. Similarly, the power of the full-grown work-buffalo does not come from high level nutrients. Continual crossbreeding among two of the classic Indian breeds has brought into being the superlative Nili-Ravi. We may seek productive syncretism, for example, in the beautiful Surti and Mediterranean breeds, or in the Murrah and the Mehsana, or the Kundi and the Surti. There is a thriving international trade in beef-type buffaloes and in buffalo frozen-semen. The possible relaxation of hitherto rigid quarantine regulations is a significant development which will stimulate reproduction research and encourage the international movement of buffalo germplasm.

3

Sheep Farm Management

Domestic Sheep

Domestic sheep (Ovis aries) are quadrupedal, ruminant mammals typically kept as livestock. Like all ruminants, sheep are members of the order Artiodactyla, the even-toed ungulates. Although the name "sheep" applies to many species, in everyday usage it almost always refers to Ovis aries. Numbering a little over 1 billion, domestic sheep are the most numerous species in their genus.

Sheep are most likely descended from the wild mouflon of Europe and Asia. One of the earliest animals to be domesticated for agricultural purposes, sheep are raised for fleece, meat (lamb, hogget or mutton) and milk. A sheep's wool is the most widely used of any animal, and is usually harvested by shearing. Ovine meat is called lamb when from younger animals and mutton when from older ones. Sheep continue to be important for wool and meat today, and are also occasionally raised for pelts, as dairy animals, or as model organisms for science.

Sheep husbandry is practised throughout the majority of the inhabited world, and has been fundamental to many civilizations. In the modern era, Australia, New Zealand, the southern and central South American nations, and the British Isles are most closely associated with sheep production.

Sheep-raising has a large lexicon of unique terms which vary considerably by region and dialect. Use of the word sheep began in Middle English as a derivation of the Old English word scçap; it is both the singular and plural name for the animal. A group of sheep is called a flock, herd or mob. Adult female sheep are referred to as ewes, intact males as rams or occasionally tups, castrated males as wethers, and younger sheep as lambs. Many other specific terms for the various life stages of sheep exist, generally related to lambing, shearing, and age.

Being a key animal in the history of farming, sheep have a deeply entrenched place in human culture, and find representation in much modern language and symbology. As livestock, sheep are most-often associated with pastoral, Arcadian imagery. Sheep figure in many mythologies—such as the Golden Fleece—and major religions, especially the Abrahamic traditions. In both ancient and modern religious ritual, sheep are used as sacrificial animals.

Description

Domestic sheep are relatively small ruminants, usually with a crimped hair called wool and often with horns forming a lateral spiral. Domestic sheep differ from their wild relatives and ancestors in several respects, having become uniquely neotenic as a result of man's influence. A few primitive breeds of sheep retain some of the characteristics of their wild cousins, such as short tails. Depending on breed, domestic sheep may have no horns at all (polled), or horns in both sexes (as in wild sheep), or in males only. Most horned breeds have a single pair, but a few breeds may have several.

Another trait unique to domestic sheep (as compared to wild ovines, not other livestock) is their wide variation in colour. Wild sheep are largely variations of brown hues, and variation with species is extremely limited. Colors of domestic sheep range from pure white to dark chocolate brown and even spotted or piebald. Selection for easily dyeable white fleeces began early in sheep domestication, and as white wool is a dominant trait it spread quickly. However, colored sheep do appear in many modern breeds, and may even appear as a recessive trait in white flocks. While white wool is desirable for large commercial markets, there is a niche market for colored fleeces, mostly for handspinning. The nature of the fleece varies widely among the breeds, from dense and highly crimped, to long and hair-like. There is variation of wool type and quality even among members of the same flock, so wool classing is a step in the commercial processing of the fibre.

Depending on breed, sheep show a range of heights and weights. Their rate of growth and mature weight is a heritable trait that is often selected for in breeding. Ewes typically weigh between 45 and 100 kilograms (99 and 220 lb), and rams between 45 and 160 kilograms (99 and 350 lb). Mature sheep have 32 teeth. As with other ruminants, the front teeth in the lower jaw bite against a hard, toothless pad in the upper jaw. These are used to pick off vegetation, then the rear teeth grind it before it is swallowed. There are eight lower front teeth in ruminants, but there is some disagreement as to whether these are eight incisors, or six incisors and two incisor-shaped canines. This means that the dental formula for sheep is either I:0/4 C:0/0 P:3/3 M:3/3, or I:0/3 C:0/1 P:3/3 M:3/3. There is a large toothless gap between the front "biting" teeth and the rear "grinding" teeth.

For the first few years of life it is possible to calculate the age of sheep from their front teeth, as a pair of milk teeth is replaced by larger adult teeth each year, the full set of eight adult front teeth being complete at about four years of age. The front teeth are then gradually lost as sheep age, making it harder for them to feed and hindering the health and productivity of the animal. For this reason, domestic sheep on normal pasture begin to slowly decline from four years on, and the average life expectancy of a sheep is 10 to 12 years, though some sheep may live as long as 20 years.

Sheep have good hearing, and are sensitive to noise when being handled. Sheep have horizontal slit-shaped pupils, possessing excellent peripheral vision; with visual fields of approximately 270° to 320°, sheep can see behind themselves without turning their heads.

However, sheep have poor depth perception; shadows and dips in the ground may cause sheep to baulk. In general, sheep have a tendency to move out of the dark and into well-lit areas, and prefer to move uphill when disturbed. Sheep also have an excellent sense of smell, and, like all species of their genus, have scent glands just in front of the eyes, and interdigitally on the feet. The purpose of these glands is uncertain, but those on the face may be used in breeding behaviours. The foot glands might also be related to reproduction, but alternative reasons, such as secretion of a waste product or a scent marker to help lost sheep find their flock, have also been proposed.

Sheep and goats are closely related as both are in the subfamily Caprinae. However, they are separate species, so hybrids rarely occur, and are always infertile. A hybrid of a ewe and a buck (a male goat) is called a sheep-goat hybrid (only a single such animal has been confirmed), and is not to be confused with the genetic chimera called a geep. Visual differences between sheep and goats include the beard and divided upper lip unique to goats. Sheep tails also hang down, even when short or docked, while the short tails of goats are held upwards. Sheep breeds are also often naturally polled (either in both sexes or just in the female), while naturally polled goats are rare (though many are polled artificially). Males of the two species differ in that buck goats acquire a unique and strong odor during the rut, whereas rams do not.

Breeds

The domestic sheep is a multi-purpose animal, and the more than 200 breeds now in existence were created to serve these diverse purposes. Some sources give a count of a thousand or more breeds, but these numbers cannot be verified. Almost all sheep are classified as being best suited to furnishing a certain product: wool, meat, milk, hides, or a combination in a dual-purpose breed. Other features used when classifying sheep include face colour (generally white or black), tail length, presence or lack of horns, and the topography for which the breed has been developed. This last point is especially stressed in the UK, where breeds are described as either upland (hill or mountain) or lowland breeds. A sheep may also be of a fat-tailed type, which is a dual-purpose sheep common in Africa and Asia with larger deposits of fat within and around its tail.

Breeds are also grouped based on how well they are suited to producing a certain type of breeding stock. Generally, sheep are thought to be either "ewe breeds" or "ram breeds". Ewe breeds are those that are hardy, and have good reproductive and mothering capabilities – they are for replacing breeding ewes in standing flocks. Ram breeds are selected for rapid growth and carcase quality, and are mated with ewe breeds to produce meat lambs. Lowland and upland breeds are also crossed in this fashion, with the hardy hill ewes crossed with larger, fast-growing lowland rams to produce ewes called mules, which can then be crossed with meat-type rams to produce prime market lambs. Many breeds, especially rare or primitive ones, fall into no clear category.

Breeds are categorized by the type of their wool. Fine wool breeds are those that have wool of great crimp and density, which are preferred for textiles. Most of these were derived from Merino sheep, and the breed continues to dominate the world sheep industry. Downs breeds have wool between the extremes, and are typically fast-growing meat and ram breeds with dark faces. Some major medium wool breeds, such as the Corriedale, are dual-purpose crosses of long and fine-wooled breeds and were created for high-production commercial flocks. Long wool breeds are the largest of sheep, with long wool and a slow rate of growth. Long wool sheep are most valued for crossbreeding to improve the attributes of other sheep types. For example: the American Columbia breed was developed by crossing Lincoln rams (a long wool breed) with fine-wooled Rambouillet ewes.

Coarse or carpet wool sheep are those with a medium to long length wool of characteristic coarseness. Breeds traditionally used for carpet wool show great variability, but the chief requirement is a wool that will not break down under heavy use (as would that of the finer breeds). As the demand for carpet-quality wool declines, some breeders of this type of sheep are attempting to use a few of these traditional breeds for alternative purposes. Others have always been primarily meat-class sheep.

A minor class of sheep are the dairy breeds. Dual-purpose breeds that may primarily be meat or wool sheep are often used secondarily as milking animals, but there are a few breeds that are predominantly used for milking. These sheep do produce a higher quantity of milk and have slightly longer lactation curves. In the quality of their milk, fat and protein content percentages of dairy sheep vary from non-dairy breeds but lactose content does not.

A last group of sheep breeds is that of fur or hair sheep, which do not grow wool at all. Hair sheep are similar to the early domesticated sheep kept before woolly breeds were developed, and are raised for meat and pelts. Some modern breeds of hair sheep, such as the Dorper, result from crosses between wool and hair breeds. For meat and hide producers, hair sheep are cheaper to keep, as they do not need shearing. Hair sheep are also more resistant to parasites and hot weather.

With the modern rise of corporate agribusiness and the decline of localized family farms, many breeds of sheep are in danger of extinction. The Rare Breeds Survival Trust of the UK lists 22 native breeds as having only 3,000 registered animals (each), and the American Livestock Breeds Conservancy lists 14 as having fewer than 10,000. Preferences for breeds with uniform characteristics and fast growth have pushed heritage (or heirloom) breeds to the margins of the sheep industry. Those that remain are maintained through the efforts of conservation organizations, breed registries, and individual farmers dedicated to their preservation.

Diet

Sheep are exclusively herbivorous mammals. Like all ruminants, sheep have a complex digestive system composed of four chambers, allowing them to break down cellulose from stems, leaves, and seed hulls into simpler carbohydrates. When sheep graze, vegetation is chewed into a mass called a bolus, which is then passed into the first chamber: the rumen. The rumen is a 19 to 38-liter (5 to 10 gal) organ in which feed is fermented via a symbiotic relationship with the bacteria, protozoa, and yeasts of the gut flora. The bolus is periodically regurgitated back to the mouth as cud for additional chewing and salivation. Cud chewing is an adaptation allowing ruminants to graze more quickly in the morning, and then fully chew and digest feed later in the day. This is beneficial as grazing, which requires lowering the head, leaves sheep vulnerable to predators, while cud chewing does not.

During fermentation, the rumen produces gas that must be expelled; disturbances of the organ, such as sudden changes in a sheep's diet, can cause potentially fatal conditions such as bloat. After fermentation in the rumen, feed passes in to the reticulum and the omasum; special feeds such as grains may bypass the rumen altogether. After the first three chambers, food moves in to the abomasum for final digestion before processing by the intestines. The abomasum is the only one of the four chambers analogous to the human stomach (being the only one that absorbs nutrients for use as energy), and is sometimes called the "true stomach".

Sheep follow a diurnal pattern of activity, feeding from dawn to dusk, stopping sporadically to rest and chew their cud. Ideal pasture for sheep is not lawn-like grass, but an array of grasses, legumes and forbs. Types of land where sheep are raised vary widely, from pastures that are seeded and improved intentionally to rough, native lands. Common plants toxic to sheep are present in most of the world, and include (but are not limited to) oak and acorns, tomato, yew, rhubarb, potato, and rhododendron.

Sheep are largely grazing herbivores, unlike browsing animals such as goats and deer that prefer taller foliage. With a much narrower face, sheep crop plants very close to the ground and can overgraze a pasture much faster than cattle. For this reason, many shepherds use managed intensive rotational grazing, where a flock is rotated through multiple pastures, giving plants time to recover. Paradoxically, sheep can both cause and solve the spread of invasive plant species. By disturbing the natural state of pasture, sheep and other livestock can pave the way for invasive plants. However, sheep also prefer to eat invasives such as cheatgrass, leafy spurge, kudzu and spotted knapweed over native species such as sagebrush, making grazing sheep effective for conservation grazing. Research conducted in Imperial County, California compared lamb grazing with herbicides for weed control in seedling alfalfa fields. Three trials demonstrated that grazing lambs were just as effective as herbicides in controlling winter weeds. Entomologists also compared grazing lambs to insecticides for insect control in winter alfalfa. In this trial, lambs provided insect control as effectively as insecticides.

Other than forage, the other staple feed for sheep is hay, often during the winter months. The ability to thrive solely on pasture (even without hay) varies with breed, but all sheep can survive on this diet. Also included in some sheep's diets are minerals, either in a trace mix or in licks.

Naturally, a constant source of potable water is also a fundamental requirement for sheep. The amount of water needed by sheep fluctuates with the season and the type and quality of the food they consume. When sheep feed on large amounts of new growth and there is precipitation (including dew, as sheep are dawn feeders), sheep need less water. When sheep are confined or are eating large amounts of cured hay, more water is typically needed. Sheep also require clean water, and may refuse to drink water that is covered in scum or algae.

Sheep are one of the few livestock animals raised for meat today that have never been widely raised in an intensive, confined animal feeding operation (CAFO). Although there is a growing movement advocating alternative farming styles, a large percentage of beef cattle, pigs, and poultry are still produced under such conditions. In contrast, only some sheep are regularly given high-concentration grain feed, much less kept in confinement. Especially in industrialized countries, sheep producers may fatten market lambs before slaughter (called "finishing") in feedlots. Many sheep breeders flush ewes and rams with a daily ration of grain during breeding to increase fertility. Ewes are also flushed during pregnancy to increase birth weights, as 70% of a lamb's growth occurs in the last five to six weeks of gestation. Otherwise, only lactating ewes and especially old or infirm sheep are commonly provided with grain. Feed provided to sheep must be specially formulated, as most cattle, poultry, pig, and even some goat feeds contain levels of copper that are lethal to sheep. The same danger applies to mineral supplements such as salt licks.

Behaviour and Intelligence

Sheep are prey animals with a strong gregarious instinct, and a majority of sheep behaviours can be understood in these terms. The dominance hierarchy of Ovis aries and its natural inclination to follow a leader to new pastures were the pivotal factors in it being one of the first domesticated livestock species. All sheep have a tendency to congregate close to other members of a flock, although this behaviour varies with breed. Farmers exploit this behaviour to keep sheep together on unfenced pastures and to move them more easily. Shepherds may also use Herding dogs in this effort, whose highly bred herding ability can assist in moving flocks. Sheep are also extremely food-oriented, and association of humans with regular feeding often results in sheep soliciting people for food. Those who are moving sheep may exploit this behaviour by leading sheep with buckets of feed, rather than forcing their movements with herding.

In regions where sheep have no natural predators, none of the native breeds of sheep exhibit a strong flocking behaviour. Sheep can also become hefted to one particular local pasture (heft) so they do not roam freely in unfenced landscapes. Ewes teach the heft to their lambs, and if whole flocks are culled it must be retaught to the replacement animals.

Flock dynamics in sheep are, as a rule, only exhibited in a group of four or more sheep. Fewer sheep may not react as normally expected when alone or with few other sheep. For sheep, the primary defense mechanism is simply to flee from danger when their flight zone is crossed. Secondly, cornered sheep may charge or threaten to do so through hoof stamping and aggressive posture. This is particularly true for ewes with newborn lambs.

In displaying flocking, sheep have a strong lead-follow tendency, and a leader often as not is simply the first sheep to move. However, sheep do establish a pecking order through physical displays of dominance. Dominant animals are inclined to be more aggressive with other sheep, and usually feed first at troughs. Primarily among rams, horn size is a factor in the flock hierarchy. Rams with different size horns may be less inclined to fight to establish pecking order, while rams with similarly sized horns are more so.

Sheep can become stressed when separated from their flock members. Sheep can recognize individual human and ovine faces, and remember them for years. Relationships in flocks tend to be closest among related sheep: in mixed-breed flocks same-breed subgroups tend to form, and a ewe and her direct descendants often move as a unit within large flocks.

Sheep are frequently thought of as extremely unintelligent animals. A sheep's herd mentality and quickness to flee and panic in the face of stress often make shepherding a difficult endeavour for the uninitiated. Despite these perceptions, a University of Illinois monograph on sheep found them to be just below pigs and on par with cattle in IQ, and some sheep have shown problem-solving abilities; a flock in West Yorkshire, England allegedly found a way to get over cattle grids by rolling on their backs, although documentation of this has relied on anecdotal accounts. In addition to long-term facial recognition of individuals, sheep can also differentiate emotional states through facial characteristics. If worked with patiently, sheep may learn their names, and many sheep are trained to be led by halter for showing and other purposes. Sheep have also responded well to clicker training. Very rarely, sheep are used as pack animals. Tibetan nomads distribute baggage equally throughout a flock as it is herded between living sites.

Reproduction

Sheep follow a similar reproductive strategy to other herd animals. A group of ewes is generally mated by a single ram, who has either been chosen by a breeder or has established dominance through physical contest with other rams (in feral populations). Most sheep are seasonal breeders, although some are able to breed year-round. Ewes generally reach sexual maturity at six to eight months of age, and rams generally at four to six months. Ewes have estrus cycles about every 17 days, during which they emit a scent and indicate readiness through physical displays towards rams. A minority of sheep display a preference for homosexuality (8% on average) or are freemartins (female animals that are behaviourally masculine and lack functioning ovaries).

Without human intervention, rams fight during the rut to determine which individuals may mate with ewes. Rams, especially unfamiliar ones, will also fight outside the breeding period to establish dominance; rams can kill one another if allowed to mix freely. During the rut, even normally friendly rams may become aggressive towards humans due to increases in their hormone levels.

After mating, sheep have a gestation period of about five months, and normal labour take one to three hours. Although some breeds regularly throw larger litters of lambs, most produce single or twin lambs. During or soon after labour, ewes and lambs may be confined to small lambing jugs, small pens designed to aid both careful observation of ewes and to cement the bond between them and their lambs.

Ovine obstetrics can be problematic. By selectively breeding ewes that produce multiple offspring with higher birth weights for generations, sheep producers have inadvertently caused some domestic sheep to have difficulty lambing; balancing ease of lambing with high productivity is one of the dilemmas of sheep breeding. In the case of any such problems, those present at lambing may assist the ewe by extracting or repositioning lambs. After the birth, ewes ideally break the amniotic sac (if it is not broken during labour), and begin licking clean the lamb. Most lambs will begin standing within an hour of birth. In normal situations, lambs nurse after standing, receiving vital colostrum milk. Lambs that either fail to nurse or that are rejected by the ewe require aid to live, such as bottle-feeding or fostering by another ewe.

After lambs are several weeks old, lamb marking (the process of ear tagging, docking, and castrating) is carried out. Vaccinations are usually carried out at this point as well. Ear tags with numbers are attached, or ear marks are applied for ease of later identification of sheep. Castration is performed on ram lambs not intended for breeding, although some shepherds choose to avoid the procedure for ethical, economic or practical reasons. Ram lambs that will either be slaughtered or separated from ewes before sexual maturity are not usually castrated. Docking, which is the shortening of a lamb's tail, is practised for health reasons. Objections to all these procedures have been raised by animal rights groups, but farmers defend them by saying they solve many practical and veterinary problems, and inflict only temporary pain.

Health

Sheep may fall victim to poisons, infectious diseases, and physical injuries. As a prey species, a sheep's system is adapted to hide the obvious signs of illness, to prevent being targeted by predators. However, there are some obvious signs of ill health, with sick sheep eating little, vocalizing excessively, and being generally listless. Throughout history, much of the money and labour of sheep husbandry has aimed to prevent sheep ailments. Historically, shepherds often created remedies by experimentation on the farm. In some developed countries, including the United States, sheep lack the economic importance for drugs companies to perform expensive clinical trials required to approve drugs for ovine use. In such instances, shepherds resort to illegal, extra-label usage of drugs approved for other animals. In the 20th and 21st centuries, a minority of sheep owners have turned to alternative treatments such as homeopathy, herbalism and even traditional Chinese medicine to treat sheep veterinary problems. Despite some favourable anecdotal evidence, the effectiveness of alternative veterinary medicine has been met with skepticism in scientific journals. The need for traditional anti-parasite drugs and antibiotics is widespread, and is the main impediment to certified organic farming with sheep.

Many breeders take a variety of preventative measures to warn off problems. The first is to ensure that all sheep are healthy when purchased. Many buyers avoid outlets known to be clearing houses for animals culled from healthy flocks as either sick or simply inferior. This can also mean maintaining a closed flock, and quarantining new sheep for a month. Two fundamental preventative programs are maintaining good nutrition and reducing stress in the sheep. Handling sheep in loud, erratic ways causes them to produce cortisol, a stress hormone. This can lead to a weakened immune system, thus making sheep far more vulnerable to disease. Signs of stress in sheep include: excessive panting, teeth grinding, restless movement, wool eating, and wood chewing. Avoiding poisoning is also important, common poisons are pesticide sprays, inorganic fertilizer, motor oil, as well as radiator coolant (the ethylene glycol antifreeze is sweet-tasting).

Common forms of preventive medication for sheep are vaccinations and treatments for parasites. Both external and internal parasites are the most prevalent malady in sheep, and are either fatal, or reduce the productivity of flocks. Worms are the most common internal parasites. They are ingested during grazing, incubate within the sheep, and are expelled through the digestive system (beginning the cycle again).

Oral anti-parasitic medicines known as drenches are given to a flock to treat worms, sometimes after worm eggs in the feces has been counted to assess infestation levels. Afterwards, sheep may be moved to a new pasture to avoid ingesting the same parasites. External sheep parasites include: lice (for different parts of the body), sheep keds, nose bots, sheep itch mite, and maggots. Keds are blood-sucking parasites that cause general malnutrition and decreased productivity, but are not fatal. Maggots are those of the bot fly and the blowfly. Fly maggots cause the extremely destructive condition of flystrike. Flies lay their eggs in wounds or wet, manure-soiled wool, when the maggots hatch they burrow into a sheep's flesh, eventually causing death if untreated. In addition to other treatments, crutching (shearing wool from a sheep's rump) is a common preventative method. Nose bots are flies that inhabit a sheep's sinuses, causing breathing difficulties and discomfort. Common signs are a discharge from the nasal passage, sneezing, and frantic movement such as head shaking. External parasites may be controlled through the use of backliners, sprays or immersive sheep dips.

A wide array of bacterial diseases affect sheep. Diseases of the hoof, such as foot rot and foot scald may occur, and are treated with footbaths and other remedies. These painful conditions cause lameness and hinder feeding. Ovine Johne's disease is a wasting disease that affects young sheep. Bluetongue disease is an insect-borne illness causing fever and inflammation of the mucous membranes. Ovine rinderpest (or peste des petits ruminants) is a highly contagious and often fatal viral disease affecting sheep and goats.

A few sheep conditions are transmittable to humans. Orf (also known as scabby mouth, contagious ecthyma or soremouth) is a skin disease leaving lesions that is transmitted through skin-to-skin contact. Cutaneous anthrax is also called woolsorter's disease as the spores can be tranmitted in unwashed wool. More seriously, the organisms that can cause spontaneous enzootic abortion in sheep are easily transmitted to pregnant women. Also of concern are the prion disease scrapie and the virus that causes foot-and-mouth disease (FMD), as both can devastate flocks. The latter poses a slight risk to humans. During the 2001 FMD pandemic in the UK, hundreds of sheep were culled and some rare British breeds were at risk of extinction due to this.

Predation

Other than parasites and disease, predation is a threat to sheep and the profitability of sheep raising. Sheep have little ability to defend themselves, compared with other species kept as livestock. Even if sheep survive an attack, they may die from their injuries, or simply from panic. However, the impact of predation varies dramatically with region. In Africa, Australia, the Americas, and parts of Europe and Asia predators are a serious problem. In the United States, for instance, over 1/3 of sheep deaths in 2004 were caused by predation. In contrast, other nations are virtually devoid of sheep predators, particularly islands known for extensive sheep husbandry. Worldwide, canids—including the domestic dog—are responsible for most sheep deaths. Other animals that occasionally prey on sheep include: felines, bears, birds of prey, ravens and feral hogs.

Sheep producers have used a wide variety of measures to combat predation. Pre-modern shepherds used their own presence, livestock guardian dogs, and protective structures such as barns and fencing. Fencing (both regular and electric), penning sheep at night and lambing indoors all continue to be widely used. More modern shepherds used guns, traps, and poisons to kill predators, causing significant decreases in predator populations. In the wake of the environmental and conservation movements, the use of these methods now usually falls under the purview of specially designated government agencies in most developed countries.

The 1970s saw a resurgence in the use of livestock guardian dogs and the development of new methods of predator control by sheep producers, many of them non-lethal. Donkeys and guard llamas have been used since the 1980s in sheep operations, using the same basic principle as livestock guardian dogs. Interspecific pasturing, usually with larger livestock such as cattle or horses, may help to deter predators, even if such species do not actively guard sheep. In addition to animal guardians, contemporary sheep operations may use non-lethal predator deterrents such as motion-activated lights and noisy alarms.

History

Sheep were among the first animals to be domesticated by humankind; sources provide a domestication date between nine and eleven thousand years ago in Mesopotamia. Their wild relatives have several characteristics—such as a relative lack of aggression, a manageable size, early sexual maturity, a social nature, and high reproduction rates—which made them particularly suitable for domestication. Today, Ovis aries is an entirely domesticated animal that is largely dependent on man for its health and survival. Feral sheep do exist, but exclusively in areas devoid of large predators (usually islands) and not on the scale of feral horses, goats, pigs, or dogs, although some feral populations have remained isolated long enough to be recognized as distinct breeds.

The exact line of descent between domestic sheep to their wild ancestors is presently unclear. The most common hypothesis states that Ovis aries is descended from the Asiatic (O. orientalis) species of mouflon. It has been proposed that the European mouflon (O. musimon) is an ancient breed of domestic sheep turned feral rather than an ancestor, despite it commonly being cited as ancestor in past literature. A few breeds of sheep, such as the Castlemilk Moorit from Scotland, were formed through crossbreeding with wild European mouflon.

The urial (O. vignei) was once thought to have been a forebear of domestic sheep, as they occasionally interbreed with mouflon in the Iranian part of their range.

However, the urial, argali (O. ammon), and snow sheep (O. nivicola) have a different number of chromosomes than other Ovis species, making a direct relationship implausible, and phylogenetic studies show no evidence of urial ancestry. Further studies comparing European and Asian breeds of sheep showed significant genetic differences between the two. Two explanations for this phenomenon have been posited. The first is that there is a currently unknown species or subspecies of wild sheep that contributed to the formation of domestic sheep. A second hypothesis suggests that this variation is the result of multiple waves of capture from wild mouflon, similar to the known development of other livestock.

Initially, sheep were kept solely for meat, milk and skins. Archaeological evidence from statuary found at sites in Iran suggests that selection for woolly sheep may have begun around 6000 BC, but the earliest woven wool garments have only been dated to two to three thousand years later. By that span of the Bronze Age, sheep with all the major features of modern breeds were widespread throughout Western Asia. However, one chief difference between ancient sheep and modern breeds is the technique by which wool could be collected. Primitive sheep cannot be shorn, and must have their wool plucked out by hand in a process called "rooing". This is because fibers called kemps are still longer than the soft fleece. The fleece may also be collected from the field after it falls out. This trait survives today in unrefined breeds such as the Soay and many Shetlands. Indeed, the Soay, along with other Northern European breeds with short tails, unshearable fleece, diminutive size, and horns in both sexes, are closely related to ancient sheep. Originally, weaving and spinning wool was a handicraft practiced at home, rather than an industry. Babylonians, Sumerians, and Persians all depended on sheep; and although linen was the first fabric to be fashioned in to clothing, wool was a prized product. The raising of flocks for their fleece was one of the earliest industries, and flocks were a medium of exchange in barter economies. Numerous biblical figures kept large flocks, and subjects of the king of Israel were taxed according to the number of rams they owned.

In Africa

Sheep entered the African continent not long after their domestication in western Asia. A minority of historians once posited a contentious African theory of origin for Ovis aries. This theory is based primarily on rock art interpretations, and osteological evidence from Barbary sheep. The first sheep entered North Africa via Sinai, and were present in ancient Egyptian society between eight and seven thousand years ago. Sheep have always been part of subsistence farming in Africa, but today the only country that keeps an influential number of commercial sheep is South Africa. South African sheep producers, in an attempt to deal with the numerous predators of Africa, invented the livestock protection collar, which holds poison at the jugular to sicken or kill predators.

In Europe

From Southwest Asia, sheep husbandry spread quickly in to Europe. Practically from its inception, ancient Greek civilization relied on sheep as primary livestock, and were even said to name individual animals. Scandinavian sheep of a type seen today — with short tails and multi-colored fleece — were also present early on. Later, the Roman Empire kept sheep on a wide scale, and the Romans were an important agent in the spread of sheep raising throughout the continent.

Pliny the Elder, in his Natural History (Naturalis Historia), speaks at length about sheep and wool. Declaring "Many thanks, too, do we owe to the sheep, both for appeasing the gods, and for giving us the use of its fleece.", he goes on to detail the breeds of ancient sheep and the many colors, lengths and qualities of wool. Romans also pioneered the practice of blanketing sheep, in which a fitted coat (today usually of nylon) is placed over the sheep to improve the cleanliness and luster of its wool.

During the Roman occupation of the British Isles, a large wool processing factory was established in Winchester, England in about 50 AD. By 1000 AD, England and Spain were recognized as the twin centres of sheep production in the Western world. As the original breeders of the fine-wooled merino sheep that have historically dominated the wool trade, the Spanish gained great wealth. Wool money largely financed Spanish rulers and thus the voyages to the New World by conquistadors. The powerful Mesta (its full title was Honrado Concejo de la Mesta, the Honorable Council of the Mesta) was a corporation of sheep owners mostly drawn from Spain's wealthy merchants, Catholic clergy and nobility that controlled the merino flocks. By the 17th century, the Mesta held in upwards of two million head of merino sheep.

Mesta flocks followed a seasonal pattern of transhumance across Spain. In the spring, they left the winter pastures (invernaderos) in Extremadura and Andalusia to graze on their summer pastures (agostaderos) in Castile, returning again in the autumn. Spanish rulers eager to increase wool profits gave extensive legal rights to the Mesta, often to the detriment of local peasantry. The huge merino flocks had a lawful right of way for their migratory routes (cañadas). Towns and villages were obliged by law to let the flocks graze on their common land, and the Mesta had its own sheriffs that could summon offending individuals to its own tribunals.

Exportation of merinos without royal permission was also a punishable offense, thus ensuring a near-absolute monopoly on the breed until the mid-18th century. After the breaking of the export ban, fine wool sheep began to be distributed worldwide. The export to Rambouillet by Louis XVI in 1786 formed the basis for the modern Rambouillet (or French Merino) breed. After the Napoleonic Wars and the global distribution of the once-exclusive Spanish stocks of Merinos, sheep raising in Spain reverted to hardy coarse-wooled breeds such as the Churra, and was no longer of international economic significance.

The sheep industry in Spain was an instance of migratory flock management, with large homogenous flocks ranging over the entire nation. The management model used in England was quite different but had a similar importance to economy of the British Empire. Up until the early 20th century, owling (the smuggling of sheep or wool out of the country) was a punishable offense, and to this day the Lord Speaker of the House of Lords sits on a cushion known as the Woolsack.

The high concentration and more sedentary nature of shepherding in the UK allowed sheep especially adapted to their particular purpose and region to be raised, thereby giving rise to an exceptional variety of breeds in relation to the land mass of the country. This greater variety of breeds also produced a valuable variety of products to compete with the superfine wool of Spanish sheep. By the time of Elizabeth I's rule, sheep and wool trade was the primary source of tax revenue to the Crown of England and the country was a major influence in the development and spread of sheep husbandry.

An important event not only in the history of domestic sheep, but of all livestock, was the work of Robert Bakewell in the 1700s. Before his time, breeding for desirable traits was often based on chance, with no scientific process for selection of breeding stock. Bakewell established the principles of selective breeding—especially line breeding—in his work with sheep, horses and cattle; his work later influenced Gregor Mendel and Charles Darwin. His most important contribution to sheep was the development of the Leicester Longwool, a quick-maturing breed of blocky conformation that formed the basis for many vital modern breeds. Today, the sheep industry in the UK has diminished significantly, though pedigreed rams can still fetch around 100,000 Pounds sterling at auction.

In the Americas

No ovine species native to the Americas has ever been domesticated, despite being closer genetically to domestic sheep than many Asian and European species. The first domestic sheep in North America—most likely of the Churra breed—arrived with Christopher Columbus' second voyage in 1493. The next transatlantic shipment to arrive was with Hernán Cortés in 1519, landing in Mexico. No export of wool or animals is known to have occurred from these populations, but flocks did disseminate throughout what is now Mexico and the Southwest United States with Spanish colonists. Churras were also introduced to the Navajo tribe of Native Americans, and became a key part of their livelihood and culture. The modern presence of the Navajo-Churro breed is a result of this heritage.

North America

The next transport of sheep to North America was not until 1607, with the voyage of the HMS Susan Conant to Virginia. However, the sheep that arrived in that year were all slaughtered because of a famine, and a permanent flock was not to reach the colony until two years later in 1609. In two decades time, the colonists had expanded their flock to a total of 400 head. By the 1640s there were about 100,000 head of sheep in the 13 colonies, and in 1662, a woolen mill was built in Watertown, Massachusetts.

Especially during the periods of political unrest and civil war in Britain spanning the 1640s and 50s which disrupted maritime trade, the colonists found it pressing to produce wool for clothing. Many islands off the coast were cleared of predators and set aside for sheep: Nantucket, Long Island, Martha's Vineyard and small islands in Boston Harbor were notable examples. There remain some rare breeds of American sheep—such as the Hog Island sheep—that were the result of island flocks. Placing semi-feral sheep and goats on islands was common practice in colonization during this period. Early on, the British government banned further export of sheep to the Americas, or wool from it, in an attempt to stifle any threat to the wool trade in the British Isles. One of many restrictive trade measures that precipitated the American Revolution, the sheep industry in the Northeast grew despite the bans.

Gradually, beginning in the 1800s, sheep production in the U.S. moved westward. Today, the vast majority of flocks reside on Western range lands. During this westward migration of the industry, competition between sheep (sometime called "range maggots") and cattle operations grew more heated, eventually erupting into range wars. Other than simple competition for grazing and water rights, cattlemen believed that the secretions of the foot glands of sheep made cattle unwilling to graze on places where sheep had stepped. As sheep production centered on the U.S. western ranges, it became associated with other parts of Western culture, such as the rodeo. In modern America, a minor event in rodeos is mutton busting, in which children compete to see who can stay atop a sheep the longest before falling off. Another effect of the westward movement of sheep flocks in North America was the decline of wild species such as Bighorn sheep (O. canadesis). Most diseases of domestic sheep are transmittable to wild ovines, and such diseases, along with overgrazing and habitat loss, are named as primary factors in the plummeting numbers of wild sheep. Sheep production peaked in North America during 1940s and 50s at more than 55 million head. Henceforth and continuing today, the number of sheep in North America has steadily declined with wool prices and the lessening American demand for sheep meat.

South America

In South America, especially in Patagonia, there is an active modern sheep industry. Sheep keeping was largely introduced through immigration to the continent by Spanish and British peoples, for whom sheep were a major industry during the period. South America has a large number of sheep, but the highest-producing nation (Brazil) kept only just over 15 million head in 2004, far fewer than most centres of sheep husbandry. The primary challenges to the sheep industry in South America are the phenomenal drop in wool prices in the late 20th century and the loss of habitat through logging and overgrazing. The most influential region internationally is that of Patagonia, which has been the first to rebound from the fall in wool prices. With few predators and almost no grazing competition (the only large native grazing mammal is the guanaco), the region is prime land for sheep raising. The most exceptional area of production is surrounding the La Plata river in the Pampas region. Sheep production in Patagonia peaked in 1952 at more than 21 million head, but has steadily fallen to fewer than ten today. Most operations focus on wool production for export from Merino and Corriedale sheep; the economic sustainability of wool flocks has fallen with the drop in prices, while the cattle industry continues to grow.

In Australia and New Zealand

Australia and New Zealand are crucial players in the contemporary sheep industry, and sheep are an iconic part of both countries' culture and economy. New Zealand has the highest density of sheep per capita (sheep outnumber the human population 12 to 1), and Australia is the world's indisputably largest exporter of sheep and cattle. In 2007, New Zealand even declared 15 February their official National Lamb Day to celebrate the country's history of sheep production.

The First Fleet brought the initial population of 70 sheep from the Cape of Good Hope to Australia in 1788. The next shipment was of 30 sheep from Calcutta and Ireland in 1793. All of the early sheep brought to Australia were exclusively used for the dietary needs of the penal colonies. The beginnings of the Australian wool industry were due to the vision and efforts of Captain John Macarthur. At Macarthur's urging 16 Spanish merinos were imported in 1797, effectively beginning the Australian sheep industry. By 1801 Macarthur had 1,000 head of sheep, and in 1803 he exported 111 kilograms (245 lb) of wool to England. Today, Macarthur is generally thought of as the father of the Australian sheep industry.

The growth of the sheep industry in Australia was explosive. In 1820, the continent held 100,000 sheep, a decade later it had one million. By 1840, New South Wales alone kept 4 million sheep; flock numbers grew to 13 million in a decade. While much of the growth in both nations was due to the active support of Britain in its desire for wool, both worked independently to develop new high-production breeds: the Corriedale, Coolalee, Coopworth, Perendale, Polwarth, Booroola Merino, Peppin Merino, and Poll Merino were all created in New Zealand or Australia. Wool production was a fitting industry for colonies far from their home nations. Before the advent of fast air and maritime shipping, wool was one of the few viable products that was not subject to spoiling on the long passage back to British ports. The abundant new land and milder winter weather of the region also aided the growth of the Australian and New Zealand sheep industries.

Flocks in Australia have always been largely range bands on fenced land, and are aimed at production of medium to superfine wool for clothing and other products as well as meat. New Zealand flocks are kept in a fashion similar to English ones, in fenced holdings without shepherds. Although wool was once the primary income source for New Zealand sheep owners (especially during the New Zealand wool boom), today it has shifted to meat production for export.

Animal Welfare Concerns

The Australian sheep industry is the only sector of the industry to receive international criticism for its practices. Sheep stations in Australia are cited in Animal Liberation, the seminal book of the animal rights movement, as the author's primary evidence in his argument against retaining sheep as a part of animal agriculture. The practice of mulesing, in which skin is cut away from an animal's perineal area to prevent cases of the fatal condition flystrike, has been condemned by PETA as being painful and unnecessary. In response, a program of phasing out mulesing is currently being implemented, and some mulesing operations are being carried out with the use of anaesthetic. The Animal Welfare Advisory Committee to the New Zealand Ministry of Agriculture Code of recommendations and minimum standards for the welfare of Sheep, considers mulesing a "special technique" which is performed on some Merino sheep at a small number of farms in New Zealand.

Most of the sheep meat exported from Australia is either frozen carcases to the UK or is live export to the Middle East. Shipped on livestock carriers in what has been called crowded, unsafe conditions by critics, live sheep are desired by Middle Eastern nations to meet the requirements of ritual halal slaughter. Opponents of the export—such as PETA—say that sheep exported to countries outside the jurisdiction of Australia's animal cruelty laws are treated with horrendous brutality and that halal facilities exist in Australia to make export of live animals redundant. A few celebrities and companies have pledged to boycott all Australian sheep products in protest.

Economic Importance

Sheep are an important part of the global agricultural economy. However, their once-vital status has been largely replaced by other livestock species, especially the pig, chicken, and cow. China, Australia, India, and Iran have the largest modern flocks, and serve both local and exportation needs for wool and mutton. Other countries such as New Zealand have smaller flocks but retain a large international economic impact due to their export of sheep products. Sheep also play a major role in many local economies, which may be niche markets focused on organic or sustainable agriculture and local food customers. Especially in developing countries, such flocks may be a part of subsistence agriculture rather than a system of trade. Sheep themselves may be a medium of trade in barter economies.

Domestic sheep provide a wide array of raw materials. Wool was one of the first textiles, although in the late 20th century wool prices began to fall dramatically as the result of the popularity and cheap prices for synthetic fabrics. For many sheep owners, the cost of shearing is greater than the possible profit from the fleece, making subsisting on wool production alone practically impossible without farm subsidies. Fleeces are used as material in making alternative products such as wool insulation. In the 21st century, the sale of meat is the most profitable enterprise in the sheep industry, even though far less sheep meat is consumed than chicken, pork or beef.

Sheepskin is likewise used for making clothes, footwear, rugs, and other products. By-products from the slaughter of sheep are also of value: sheep tallow can be used in candle and soap making, sheep bone and cartilage has been used to furnish carved items such as dice and buttons as well as rendered glue and gelatin. Sheep intestine can be formed into sausage casings, and lamb intestine has been formed into surgical sutures, as well as strings for musical instruments and tennis rackets. Sheep droppings, which are high in cellulose, have even been sterilized and mixed with traditional pulp materials to make paper. Of all sheep by-products, perhaps the most valuable is lanolin: the water-proof, fatty substance found naturally in sheep's wool and used as a base for innumerable cosmetics and other products.

Some farmers who keep sheep also make a profit from live sheep. Providing lambs for youth programs such as 4-H and competition at agricultural shows is often a dependable avenue for the sale of sheep. Farmers may also choose to focus on a particular breed of sheep in order to sell registered purebred animals, as well as provide a ram rental service for breeding. The most valuable sheep ever sold to date was a purebred Texel ram that fetched £231,000 at auction. The previous record holder was a Merino ram sold for £205,000 in 1989. A new option for deriving profit from live sheep is the rental of flocks for grazing; these "mowing services" are hired in order to keep unwanted vegetation down in public spaces and to lessen fire hazard.

Despite the falling demand and price for sheep products in many markets, sheep have distinct economic advantages when compared with other livestock. They do not require the expensive housing, such as that used in the intensive farming of chickens or pigs. They are an efficient use of land; roughly six sheep can be kept on the amount that would suffice for a single cow or horse. Sheep can also consume plants, such as noxious weeds, that most other animals will not touch, and produce more young at a faster rate. Also, in contrast to most livestock species, the cost of raising sheep is not necessarily tied to the price of feed crops such as grain, soybeans and corn. Combined with the lower cost of quality sheep, all these factors combine to equal a lower overhead for sheep producers, thus entailing a higher profitability potential for the small farmer. Sheep are especially beneficial for independent producers, including family farms with limited resources, as the sheep industry is one of the few types of animal agriculture that has not been vertically integrated by agribusiness.

As food

Sheep meat and milk were one of the earliest staple proteins consumed by human civilization after the transition from hunting and gathering to agriculture. Sheep meat prepared for food is known as either mutton or lamb. "Mutton" is derived from the Old French moton, which was the word for sheep used by the Anglo-Norman rulers of much of the British Isles in the Middle Ages. This became the name for sheep meat in English, while the Old English word sceap was kept for the live animal. Throughout modern history, "mutton" has been limited to the meat of mature sheep usually at least two years of age; "lamb" is used for that of immature sheep less than a year.

In the 21st century, the nations with the highest consumption of sheep meat are the Persian Gulf states, New Zealand, Australia, Greece, Uruguay, the United Kingdom and Ireland. These countries eat 14–40 lbs (3–18 kg) of sheep meat per capita, per annum. Sheep meat is also popular in France, Africa (especially the Maghreb), the Caribbean, the rest of the Middle East, India, and parts of China. This often reflects a past history of sheep production. In these countries in particular, dishes comprising alternative cuts and offal may be popular or traditional. Sheep testicles—called animelles or lamb fries—are considered a delicacy in many parts of the world. Perhaps the most unusual dish of sheep meat is the Scottish haggis, composed of various sheep innards cooked along with oatmeal and chopped onions inside its stomach. In comparison, countries such as the U.S. consume only a pound or less (under 0.5 kg), with Americans eating 50 pounds (22 kg) of pork and 65 pounds (29 kg) of beef. In addition, such countries rarely eat mutton, and may favour the more expensive cuts of lamb: mostly lamb chops and leg of lamb.

Though sheep's milk may be drunk rarely in fresh form, today it is used predominantly in cheese and yogurt making. Sheep have only two teats, and produce a far smaller volume of milk than cows. However, as sheep's milk contains far more fat, solids, and minerals than cow's milk, it is ideal for the cheese-making process. It also resists contamination during cooling better because of its much higher calcium content. Well-known cheeses made from sheep milk include the Feta of Bulgaria and Greece, Roquefort of France, Manchego from Spain, the Pecorino Romano (the Italian word for sheep is pecore) and Ricotta of Italy. Yogurts, especially some forms of strained yogurt, may also be made from sheep milk. Many of these products are now often made with cow's milk, especially when produced outside their country of origin. Sheep milk contains 4.8% lactose, which may affect those who are intolerant.

In Science

Sheep are generally too large and reproduce too slowly to make ideal research subjects, and thus are not a common model organism. They have, however, played an influential role in some fields of science. In particular, the Roslin Institute of

Edinburgh, Scotland used sheep for genetics research that produced groundbreaking results. In 1995, two ewes named Megan and Morag were the first mammals cloned from differentiated cells. A year later, a Finnish Dorset sheep named Dolly, dubbed "the world's most famous sheep" in Scientific American, was the first mammal to be cloned from an adult somatic cell. Following this, Polly and Molly were the first mammals to be simultaneously cloned and transgenic. As of 2008, the sheep genome has not been fully sequenced, although a detailed genetic map has been published, and a draft version of the complete genome produced by assembling sheep DNA sequences using information given by the genomes of other mammals.

In the study of natural selection, the population of Soay sheep that remain on the island of Hirta have been used to explore the relation of body size and coloration to reproductive success. Soay sheep come in several colors, and researchers investigated why the larger, darker sheep were in decline; this occurrence contradicted the rule of thumb that larger members of a population tend to be more successful reproductively. The feral Soays on Hirta are especially useful subjects because they are isolated.

Sheep are one of the few animals where the molecular basis of the diversity of male sexual preferences has been examined. However, this research has been controversial, and much publicity has been produced by a study at the Oregon Health and Science University that investigated the mechanisms that produce homosexuality in rams. Organizations such as PETA campaigned against the study, accusing scientists of trying to cure homosexuality in the sheep. OHSU and the involved scientists vehemently denied such accusations.

Domestic sheep are sometimes used in medical research, particularly for researching cardiovascular physiology, in areas such as hypertension and heart failure. Pregnant sheep are also a useful model for human pregnancy, and have been used to investigate the effects on fetal development of malnutrition and hypoxia. In behavioural sciences, sheep have been used in isolated cases for the study of facial recognition, as their mental process of recognition is qualitatively similar to humans.

Cultural Impact

Sheep have had a strong presence in many cultures, especially in areas where they form the most common type of livestock. In the English language, to call someone a sheep or ovine may allude that they are timid and easily led, if not outright stupid. In contradiction to this image, male sheep are often used as symbols of virility and power, such as for the St. Louis Rams and the Dodge Ram. Sheep are key symbols in fables and nursery rhymes like The Wolf in Sheep's Clothing, Little Bo Peep, Baa, Baa, Black Sheep, and Mary Had a Little Lamb. Novels such as George Orwell's Animal Farm, Haruki Murakami's A Wild Sheep Chase, Thomas Hardy's Far from the Madding Crowd and Three Bags Full: A Sheep Detective Story utilize sheep as characters or plot devices. Poems like William Blake's "The Lamb", songs such as Pink Floyd's Sheep and Bach's aria Sheep may safely graze (Schafe können sicher weiden) use sheep for metaphorical purposes. In more recent popular culture, the 2007 film Black Sheep exploits sheep for horror and comedic effect, ironically turning them into blood-thirsty killers.

Counting sheep is popularly said to be an aid to sleep, and some ancient systems of counting sheep persist today. Sheep also enter in colloquial sayings and idiom frequently with such phrases as "black sheep". To call an individual a black sheep implies that they are an odd or disreputable member of a group. This usage derives from the recessive trait that causes an occasional black lamb to be born in to an entirely white flock. These black sheep were considered undesirable by shepherds, as black wool is not as commercially viable as white wool. Citizens who accept overbearing governments have been referred to by the Portmanteau neologism of sheeple. Somewhat differently, the adjective "sheepish" is also used to describe embarrassment.

In Religion and Folklore

In antiquity, symbolism involving sheep cropped up in religions in the ancient Near East, the Mideast, and the Mediterranean area: Çatalhöyük, ancient Egyptian religion, the Cana'anite and Phoenician tradition, Judaism, Greek religion, and others. Religious symbolism and ritual involving sheep began with some of the first known faiths: skulls of rams (along with bulls) occupied central placement in shrines at the Çatalhöyük settlement in 8,000 BCE. In Ancient Egyptian religion, the ram was the symbol of several gods: Khnum, Heryshaf and Amun (in his incarnation as a god of fertility). Other deities occasionally shown with ram features include: the goddess Ishtar, the Phoenician god Baal-Hamon, and the Babylonian god Ea-Oannes. In Madagascar, sheep were not eaten as they were believed to be incarnations of the souls of ancestors.

There are also many ancient Greek references to sheep: that of Chrysomallos, the golden-fleeced ram, continuing to be told through into the modern era. Astrologically, Aries, the ram, is the first sign of the classical Greek zodiac and the sheep is also the eighth of the twelve animals associated with the 12-year cycle of in the Chinese zodiac, related to the Chinese calendar. In Mongolia, shagai are an ancient form of dice made from the cuboid bones of sheep that are often used for fortunetelling purposes.

Sheep play an important role in all the Abrahamic faiths; Abraham, Isaac, Jacob, Moses, Muhammad and King David were all shepherds. According to the story of the Binding of Isaac, a ram is sacrificed as a substitute for Isaac after an angel stays Abraham's hand. Eid al-Adha is a major annual festival in Islam in which sheep (or other animals) are sacrificed in remembrance of this act. Sheep are also occasionally sacrificed to commemorate important secular events in Islamic cultures. Greeks and Romans also sacrificed sheep regularly in religious practice, and Judaism also once sacrificed sheep as a Korban (sacrifice), such as the Passover lamb. Ovine symbols—such as the ceremonial blowing of a shofar—still find a presence in modern Judaic traditions. Followers of Christianity are collectively often referred to as a flock, with Christ as the Good Shepherd, and sheep are an element in the Christian iconography of the birth of Jesus. Some Christian saints are considered patrons of shepherds, and even of sheep themselves. Christ is also portrayed as the Sacrificial lamb of God (Agnus Dei) and Easter celebrations in Greece and Romania traditionally feature a meal of Paschal lamb.

Goat

The domestic goat (Capra aegagrus hircus) is a subspecies of goat domesticated from the wild goat of southwest Asia and Eastern Europe. The goat is a member of the Bovidae family and is closely related to the sheep: both are in the goat-antelope subfamily Caprinae. There are over three hundred distinct breeds of goat.

Goats are one of the oldest domesticated species. Goats have been used for their milk, meat, hair, and skins over much of the world. In the twentieth century they also gained in popularity as pets.

Female goats are referred to as does or nannies, intact males as bucks or billies; their offspring are kids. Note that many goat breeders prefer the terms "buck" and "doe" to "billy" and "nanny". Castrated males are wethers. Goat meat from younger animals is called kid or cabrito, and from older animals is sometimes called chevon, or in some areas "mutton".

History

The most recent genetic analysis confirms the archaeological evidence that the Anatolian Zagros are the likely origin of almost all domestic goats today. Neolithic farmers began to keep them for easy access to milk and meat, primarily, also for their dung, which was used as fuel and their bones, hair, and sinew for clothing, building, and tools. The earliest remnants of domesticated goats dating 10,000 before present are found in Ganj Dareh in Iranian Kurdistan. Domestic goats were generally kept in herds that wandered on hills or other grazing areas, often tended by goatherds who were frequently children or adolescents, similar to the more widely known shepherd. These methods of herding are still used today.

Historically, goat hide has been used for water and wine bottles in both travelling and transporting wine for sale. It has also been used to produce parchment.

Anatomy

Most goats naturally have two horns, of various shapes and sizes depending on the breed. While horns are a predominantly male feature, some breeds of goat have horned females. Polled (hornless) goats are not uncommon and there have been incidents of polycerate goats (having as many as eight horns), although this is a genetic rarity thought to be inherited. Their horns are made of living bone surrounded by keratin and other proteins, and are used for defense, dominance, and territoriality.

Goats are ruminants. They have a four-chambered stomach consisting of the rumen, the reticulum, the omasum, and the abomasum.

Goats have horizontal slit-shaped pupils, an adaptation which increases peripheral depth perception. Because goats' irises are usually pale, the pupils are much more visible than in animals with horizontal pupils, but very dark irises, such as cattle, deer, most horses and many sheep.

Both male and female goats have beards, and many types of goat (most commonly dairy goats, dairy-cross boers, and pygmy goats) may have wattles, one dangling from each side of the neck.

Some breeds of sheep and goats look similar, but they can usually be told apart because goat tails are short and point up, whereas sheep tails hang down and are usually longer and bigger – though some (like those of Northern European short-tailed sheep) are short, and longer ones are often docked.

Reproduction

In some climates, goats are able to breed at any time of the year. In temperate climates and among the Swiss breeds, the breeding season commences as the day length shortens, and ends in early spring. Does of any breed come into heat every 21 days for 2 to 48 hours. A doe in heat typically flags her tail often, stays near the buck if one is present, becomes more vocal, and may also show a decrease in appetite and milk production for the duration of the heat.

Bucks (intact males) of Swiss and northern breeds come into rut in the fall as with the doe's heat cycles. Rut is characterized by a decrease in appetite and obsessive interest in the does.

In addition to natural mating, artificial insemination has gained popularity among goat breeders, as it allows easy access to a wide variety of bloodlines.

Gestation length is approximately 150 days. Twins are the usual result, with single and triplet births also common. Less frequent are litters of quadruplet, quintuplet, and even sextuplet kids. Birthing, known as kidding, generally occurs uneventfully. Right before kidding the doe will have a sunken area around the tail and hip. Also she will have heavy breathing, a worried look, become restless and show great display of affection for her keeper. The mother often eats the placenta, which gives her much needed nutrients, helps stanch her bleeding, and parallels the behaviour of wild herbivores such as deer to reduce the lure of the birth scent for predators.

Freshening (coming into milk production) occurs at kidding. Milk production varies with the breed, age, quality, and diet of the doe; dairy goats generally produce between 660 to 1,800 L (1,500 and 4,000 lb) of milk per 305 day lactation. On average, a good quality dairy doe will give at least 6 lb (2.7 l) of milk per day while she is in milk. A first time milker may produce less, or as much as 16 lb (7.3 l), or more of milk in exceptional cases. After the 305 day lactation, the doe will "dry off", typically after she has been bred. Occasionally, goats that have not been bred and are continuously milked will continue lactation beyond the typical 305 days. Meat, fibre, and pet breeds are not usually milked and simply produce enough for the kids until weaning.

Diet

Goats are reputed to be willing to eat almost anything, except tin cans and cardboard boxes. While goats will not actually eat inedible material, they are browsing animals, not grazers like cattle and sheep, and (coupled with their natural curiosity) will chew on and taste just about anything resembling plant matter in order to decide whether it is good to eat, including cardboard and paper labels from tin cans. Another possibility is that the goats are curious about the unusual smells of leftover food in discarded cans or boxes.

Aside from sampling many things, goats are quite particular in what they actually consume, preferring to browse on the tips of woody shrubs and trees, as well as the occasional broad-leaved plant. However, it can fairly be said that their plant diet is extremely varied, and includes some species which are otherwise toxic. They will seldom consume soiled food or contaminated water unless facing starvation. This is one of the reasons why goat rearing is most often free ranging, since stall-fed goat rearing involves extensive upkeep and is seldom commercially viable.

Goats prefer to browse on shrubbery and weeds, more like deer than sheep, preferring them to grasses. Nightshade is poisonous; wilted fruit tree leaves can also kill goats. Silage (corn stalks) is not good for goats, but haylage can be used if consumed immediately after opening. Alfalfa is their favorite hay; fescue is the least palatable and least nutritious. Mold in a goat's feed can make it sick and possibly kill it. Goats should not be fed grass with any signs of mold.

The digestive physiology of a very young kid (like the young of other ruminants) is essentially the same as that of a monogastric animal. Milk digestion begins in the abomasum, the milk having bypassed the rumen via closure of the reticular/esophageal groove during suckling. At birth, the rumen is undeveloped, but as the kid begins to consume solid feed, the rumen soon increases in size and in its capacity to absorb nutrients.

Behaviour

Goats are extremely curious and intelligent. They are easily trained to pull carts and walk on leads. Ches McCartney, nicknamed "the goat man", toured the United States for over three decades in a wagon pulled by a herd of pet goats. They are also known for escaping their pens. Goats will test fences, either intentionally or simply because they are handy to climb on. If any of the fencing can be spread, pushed over or down, or otherwise be overcome, the goats will escape. Being very intelligent, once a weakness in the fence has been discovered, it will be exploited repeatedly. Goats are very coordinated and can climb and hold their balance in the most precarious places. Goats are also widely known for their ability to climb trees, although the tree generally has to be on somewhat of an angle. The vocalization goats make is called bleating.

Goats have an intensely inquisitive and intelligent nature: they will explore anything new or unfamiliar in their surroundings. They do so primarily with their prehensile upper lip and tongue. This is why they investigate items such as buttons, camera cases or clothing (and many other things besides) by nibbling at them, occasionally even eating them.

Life Expectancy

Life expectancy for goats is between 15 and 18 years. An instance of a goat reaching the age of 24 has been reported.

Several factors can reduce this average expectancy, however; problems during kidding can lower a doe's expected life span to 10 or 11, and stresses of going into rut can lower a buck's expected life span to 8 or 10.

Goats in Agriculture

A goat is useful to humans either living or dead, first as a renewable provider of milk and fiber, and then as meat and hide. Some charities provide goats to impoverished people in poor countries, because goats are easier and cheaper to manage than cattle, and have multiple uses. In addition, goats are used for driving and packing purposes.

For instance, the intestine is used to make "catgut", which is still in use as a material for internal human surgical sutures and strings for musical instruments. The horn of the goat, which signifies wellbeing (Cornucopia), is also used to make spoons.

Meat

The taste of goat meat is similar to that of lamb meat; in fact, in the English-speaking islands of the Caribbean, and in some parts of Asia, particularly Pakistan and India, the word "mutton" is used to describe both goat and lamb meat. However, some feel that it has a similar taste to veal or venison, depending on the age and condition of the goat. The flavor of goat meat is said to be primarily linked to the presence of 4-methyloctanoic and 4-methylnonanoic acid. It can be prepared in a variety of ways including stewed, baked, grilled, barbecued, minced, canned, fried, curried, or made into sausage. Due to its low fat content, the meat can toughen at high temperatures without additional moisture. One of the most popular goats grown for meat is the South African Boer, introduced into the United States in the early 1990s. The New Zealand Kiko is also considered a meat breed, as is the myotonic or "fainting goat", a breed originating in Tennessee.

Milk, Butter and Cheese

Goats produce approximately 2% of the world's total annual milk supply. Goat milk differs from cow or human milk by having higher digestibility, distinct alkalinity, higher buffering capacity, and certain therapeutic values in human medicine and nutrition.

Basic Composition of Various Milks (Mean Values per 100g)

Constituent	Goat	Cow	Human
Fat (g)	3.8	3.6	4.0
Protein (g)	3.5	3.3	1.2
Lactose (g)	4.1	4.6	6.9
Ash (g)	0.8	0.7	0.2
Total solids (g)	12.2	12.3	12.3
Calories	70	69	68

Some goats are bred for milk, which can be drunk raw, although some people recommend pasteurization to reduce bacteria such as Staphylococcus aureus and Escherichia coli. If the strong-smelling buck is not separated from the does, his scent will affect the milk. Goat's milk is commonly processed into cheese, goat butter, ice cream, cajeta and other products.

Goat's milk can replace sheep's milk or cow's milk in diets of those who are allergic. However, like cow's milk, goat's milk has lactose (sugar), and may cause gastrointestinal problems for individuals with lactose intolerance.

Goat's milk naturally has small fat globules, which means the cream remains suspended in the milk, instead of rising to the top, as in raw cow's milk; therefore, it does not need to be homogenized.

Many dairy goats, in their prime, average 6 to 8 pounds (2.7 to 3.6 kg) of milk daily (roughly 3 to 4 US quarts (2.7 to 3.6 liters)) during a ten-month lactation, producing more after freshening and gradually dropping in production toward the end of their lactation. The milk generally averages 3.5 percent butterfat. A doe may be expected to reach her heaviest production during her third or fourth lactation. It is also said that "formula derived from goat's milk is unsuitable for babies who are lactose intolerant as it contains levels of lactose similar to cow's-milk-based infant formulae."

Goat butter is white because goats produce milk with the yellow beta-carotene converted to a colourless form of vitamin A.

Goat cheese is known as chèvre in France, after the French word for "goat". Some varieties include Rocamadour and Montrachet.

Fiber

The Angora breed of goats produces long, curling, lustrous locks of mohair. The entire body of the goat is covered with mohair and there are no guard hairs. The locks constantly grow and can be four inches or more in length. Angora crossbreeds, such as the pygora and the nigora, have been created to produce mohair and/or cashgora on a smaller, easier-to-manage animal. The wool is shorn (cut from the body) twice a year, with an average yield of about 10 pounds.

Most goats have softer insulating hairs nearer the skin, and longer guard hairs on the surface. The desirable fiber for the textile industry is the former, and it goes by several names (down, cashmere and pashmina). The coarse guard hairs are of little value as they are too coarse, difficult to spin and difficult to dye. The cashmere goat produces a commercial quantity of cashmere wool, which is one of the most expensive natural fibers commercially produced; cashmere is very fine and soft. The cashmere goat fiber is harvested once a year, yielding around 9 ounces (200 grammes) of down.

In South Asia, cashmere is called "pashmina" (from Persian pashmina, "fine wool"). In the 18th and early 19th century, Kashmir (then called Cashmere by the English), had a thriving industry producing shawls from goat down imported from Tibet and Tartary through Ladakh. The shawls were introduced into Western Europe when the General in Chief of the French campaign in Egypt (1799-1802) sent one to Paris. Since these shawls were produced in the upper Kashmir and Ladakh region, the wool came to be known as "cashmere".

Goat Breeds

Goat breeds fall into overlapping, general categories. They are generally distributed in to those used for dairy, fiber, meat, skins, and as companion animals. Some breeds are also particularly noted as pack goats.

Showing

Goat breeders' clubs frequently hold shows, where goats are judged on traits relating to conformation, udder quality, evidence of high production, longevity, build and muscling (meat goats and pet goats) and fiber production and the fiber itself (fiber goats). People who show their goats usually keep registered stock and the offspring of award-winning animals command a higher price. Registered goats, in general, are usually higher-priced if for no other reason than that records have been kept proving their ancestry and the production and other data of their sires, dams, and other ancestors. A registered doe is usually less of a gamble than buying a doe at random (as at an auction or sale barn) because of these records and the reputation of the breeder. Children's clubs such as 4-H also allow goats to be shown. Children's shows often include a showmanship class, where the cleanliness and presentation of both the animal and the exhibitor as well as the handler's ability and skill in handling the goat are scored. In a showmanship class, conformation is irrelevant since this is not what is being judged.

Various "Dairy Goat Scorecards" (milking does) are systems used for judging shows in the US. The American Dairy Goat Association (ADGA) scorecard for an adult doe includes a point system of a hundred total with major categories that include general appearance, the dairy character of a doe (physical traits that aid and increase milk production), body capacity, and specifically for the mammary system. Young stock and bucks are judged by different scorecards which place more emphasis on the other three categories; general appearance, body capacity, and dairy character.

The American Goat Society (AGS) has a similar, but not identical scorecard that is used in their shows. The miniature dairy goats may be judged by either of the two scorecards. The "Angora Goat scorecard" used by the Colored Angora Goat Breeder's Association CAGBA (which covers the white and the colored goats) includes evaluation of an animal's fleece colour, density, uniformity, fineness, and general body confirmation. Disqualifications include: a deformed mouth, broken down pasterns, deformed feet, crooked legs, abnormalities of testicles, missing testicles, more than 3 inch split in scrotum, and close-set or distorted horns.

Religion, Mythology and Folklore

According to Norse mythology, the god of thunder, Thor, has a chariot that is pulled by the goats Tanngrisnir and Tanngnjóstr. At night when he sets up camp, Thor eats the meat of the goats, but take care that all bones remain whole. Then he wraps the remains up, and in the morning, the goats always come back to life to pull the chariot. When a farmer's son who is invited to share the meal breaks one of the goats' leg bones to suck the marrow, the animal's leg remains broken in the morning, and the boy is forced to serve Thor as a servant to compensate for the damage.

Possibly related, the Yule Goat is one of the oldest Scandinavian and Northern European Yule and Christmas symbols and traditions. Yule Goat originally denoted the goat that was slaughtered around Yule, but it may also indicate a goat figure made out of straw. It is also used about the custom of going door-to-door singing carols and getting food and drinks in return, often fruit, cakes and sweets. "Going Yule Goat" is similar to the British custom wassailing, both with heathen roots. The Gävle Goat is a giant version of the Yule Goat, erected every year in the Swedish city of Gävle.

The Greek god, Pan, is said to have the upper body of a man and the horns and lower body of a goat. Pan was a very lustful god, nearly all of the myths involving him had to do with him chasing nymphs. He is also credited with creating the pan flute.

The goat is one of the twelve-year cycle of animals which appear in the Chinese zodiac related to the Chinese calendar. Each animal is associated with certain personality traits; those born in a year of the goat are predicted to be shy, introverted, creative, and perfectionist.

Several mythological hybrid creatures are believed to consist of parts of the goat, including the Chimera. The Capricorn sign in the Western zodiac is usually depicted as a goat with a fish's tail. Fauns and satyrs are mythological creatures that are part goat and part human. The mineral bromine is named from the Greek word "brümos," which means "stench of he-goats."

Goats are mentioned many times in the Bible. A goat is considered a "clean" animal by Jewish dietary laws and was slaughtered for an honored guest. It was also acceptable for some kinds of sacrifices. Goat-hair curtains were used in the tent that contained the tabernacle (Exodus 25:4). On Yom Kippur, the festival of the Day of Atonement, two goats were chosen and lots were drawn for them. One was sacrificed and the other allowed to escape into the wilderness, symbolically carrying with it the sins of the community. From this comes the word "scapegoat". A leader or king was sometimes compared to a male goat leading the flock. In the New Testament, Jesus told a parable of The Sheep and the Goats.

Christianity has associated Satan with imagery of goats). A common superstition in the Middle Ages was that goats whispered lewd sentences in the ears of the saints. The origin of this belief was probably the behaviour of the buck in rut, the very epitome of lust. The common medieval depiction of the Devil was that of a goat-like face with horns and small beard (a goatee). The Black Mass, a probably-mythological "Satanic mass," was said to involve a black goat, the form in which Satan supposedly manifested himself for worship.

The goat has had a lingering connection with Satanism and pagan religions, even into modern times. The inverted pentagram, a symbol used in Satanism, is said to be shaped like a goat's head. The "Baphomet of Mendes" refers to a satanic goat-like figure from 19th century occultism.

Feral Goats

Goats readily revert to the wild if given the opportunity. They have established themselves in many areas: feral goats occur in Australia, New Zealand, Great Britain, the Galapagos and in many other places. When feral goats reach large populations in habitats which are not adapted to them, they may have serious negative effects, such as removing native scrub, trees and other vegetation. Feral goats are a severe problem in Australia. However, in other circumstances they may become a natural component of the habitat.

Goat Farming In India

Goat is known as 'Poor man's cow' in India and is a very important component in dry land farming system. Marginal or undulating lands unsuitable for other types of animals like cow or buffalo, goat is the best alternative. With very low investments goat rearing can be made in to a profitable venture for small and marginal farmers.

Who can start this?

- Small and marginal farmers
- Landless labourers
- Availability of common grazing lands.

Reasons to start:

- Low capital investment and quick return
- Simple and small shed is enough
- Profitable under stall fed condition.
- High prolific rate of goats
- Year around job
- meat is lean & has low fats and is liked by all people
- Can be sold and encashed anytime.

Which Breed is Right for You?

Jamunapari:

- Fairly tall animal
- Has a strongly arched Roman nose and long pendulous ears which are about 12 inches long in adult.Jamunapari.
- Buck weighs about 65-85 kgs and does weighs about 45-60 kgs
- one kid per kidding
- six months kid weighs about 15 kg
- Milk production is about 2-2.5 lit/day.

Tellichery:

- Goat are in white, brown and black colors
- Kidding size is 2-3 kids
- Buck weighs about 40-50 kgs and does weighs about 30 kgs.

Boer:

- Rearing for meat all over the world
- Has a fast growth rate
- Buck weighs about 110-135 kgs and does weighs about 90-100 kgs
- Kids weighs about 20-30 kgs at 90 days of age.

Selection of Goats for Breeding

Does:

- should have kidding size of 2-3 kids
- should mature at 6-9 months.

Bucks:

- Tall with broad chest and slender body
- matures at age of 9-12 months
- Select kids with good body weight at 6 months of age
- select from dam of 2-3 kidding size.

Breeding Management:

- For a profitable goat farming there should be 3 kidding for 2 years
- Goats with faster growth rate and larger size should be used for breeding.
- Does of one year of age should be used for breeding
- Does should bred after 3 months of kidding then only there will be 3 kidding for 2 years.
- Goats come into heat or estrus approximately every 18 to 21 days and the heat will last for 24-72 hours.
- Does in heat become vocal and some bleat very loudly as if in pain. Constant tail wagging from side to side is another sign of heat. In addition, the vulva will appear slightly swollen and reddened and the area around the tail may look wet and dirty because of vaginal discharge. There is decreased appetite and an increased frequency of urination. Doe in heat may mount another doe as if she were a buck or let another doe mount her.
- Does can be bred 12-18 hours after the starting the estrus symptoms
- In some does the estrous period will last for 2-3 days. So they should bred again in the next day.
- Gestation period is approximately 145-150 days, but up to one week either way is fairly common. Better to be prepared well in advance.

Deworming:

- Before breeding the doe should be dewormed. Those goats affected with worm infestation are lean and dull
- Kids should be dewormed at one month of age. The life cycle of worms is three weeks, so de-worming again at two months of age is recommended
- Pregnant does should be dewormed 2-3 weeks before kidding.
- Does in early pregnancy (upto 2 months) should not be dewormed to avoid abortion.

Vaccinations:

- Kids should be given first dose of Enterotoxaemia and Tetanus vaccination at 8 weeks of age and again at 12 weeks of age.
- Does should be given Enterotoxaemia and Tetanus vaccination 4-6 weeks before breeding season and 4-6 weeks before kidding.
- Bucks should be vaccinated with Enterotoxaemia and Tetanus once in a year.

Housing for Goats

Deep litter system:

- A small shed with good cross ventilation is enough to keep a small herd.
- Litter height should be atleat 6cms.
- Litter material maybe of sawdust, paddy husk and groundnut shell.
- The litter material has to be turned periodically to remove the foul odour in the pen.
- Litter material should be replaced once in every 2 weeks.
- Each goat requires about 15 sq.ft area.
- care should be taken to reduce the external parasitic infestation.
- An adult goat produces about a tonne of manure in a year.

Raised platform system;

- wooden plank or wire mesh is placed about 3-4 ft from the ground level
- External parasitic infestation is usually less in this method.

Raising methods:

1. Semi intensive system
   * Places where less grazing lands, goats can be fed intensively with green fodders and concentrates after grazing
2. Intensive system
   * Goats are fed with green fodder and concentrates in the shed.
   * No grazing
   * house maybe of deep litter system or raised platform system.

Goat Insurance:

- Goats can be insured from 4 months of age through General Insurance companies
- Insurance can be claimed if death due to accident or diseases.

4

Beekeeping Management

Beekeeping (or apiculture, from Latin apis, bee) is the maintenance of honey bee colonies, commonly in hives, by humans. A beekeeper (or apiarist) keeps bees in order to collect honey and beeswax, to pollinate crops, or to produce bees for sale to other beekeepers. A location where bees are kept is called an apiary.

History of Beekeeping

Origins

There are more than 20,000 species of wild bees. Many species are solitary, and many others rear their young in burrows and small colonies, like mason bees and bumblebees. Beekeeping, or apiculture, is concerned with the practical management of the social species of honey bees, which live in large colonies of up to 100,000 individuals.

In Europe and America the species universally managed by beekeepers is the Western honey bee (Apis mellifera). This species has several sub-species or regional varieties, such as the Italian bee (Apis mellifera ligustica), European dark bee (Apis mellifera mellifera), and the Carniolan honey bee (Apis mellifera carnica). In the tropics, other species of social bee are managed for honey production, including Apis cerana.

All of the Apis mellifera sub-species are capable of inter-breeding and hybridizing. Many bee breeding companies strive to selectively breed and hybridize varieties to produce desirable qualities: disease and parasite resistance, good honey production, swarming behaviour reduction, prolific breeding, and mild disposition. Some of these hybrids are marketed under specific brand names, such as the Buckfast Bee or Midnite Bee. The advantages of the initial F1 hybrids produced by these crosses include: hybrid vigor, increased honey productivity, and greater disease resistance. The disadvantage is that in subsequent generations these advantages may fade away and hybrids tend to be very defensive and aggressive.

Wild Honey Harvesting

Collecting honey from wild bee colonies is one of the most ancient human activities and is still practiced by aboriginal societies in parts of Africa, Asia, Australia, and South America. Some of the earliest evidence of gathering honey from wild colonies is from rock paintings, dating to around 13,000 BCE. Gathering honey from wild bee colonies is usually done by subduing the bees with smoke and breaking open the tree or rocks where the colony is located, often resulting in the physical destruction of the nest location.

Domestication of Wild Bees

At some point humans began to domesticate wild bees in artificial hives made from hollow logs, wooden boxes, pottery vessels, and woven straw baskets or "skeps". Honeybees were kept in Egypt from antiquity. On the walls of the sun temple of Nyuserre Ini from the 5th Dynasty, before 2422 BCE, workers are depicted blowing smoke into hives as they are removing honeycombs. Inscriptions detailing the production of honey are found on the tomb of Pabasa from the 26th Dynasty (circa 650 BCE), depicting pouring honey in jars and cylindrical hives. Sealed pots of honey were found in the grave goods of Pharaohs such as Tutankhamun.

In prehistoric Greece (Crete and Mycenae), existed a systematic of high-status apiculture as it can be concluded by the finds of hives, smoking pots, honey extractors and other beekeeping paraphernalia in Knossos. Beekeeping was considered a highly valued industry controlled by beekeeping overseers-owners of gold rings depicting apiculture scenes rather that religious ones as they have been reinterpreted recently, contra Sir Arthur Evans.

Archaeological finds relating to beekeeping have been discovered at Rehov, a Bronze-and Iron Age archaeological site in the Jordan Valley, Israel. Thirty intact hives, made of straw and unbaked clay, were discovered by archaeologist Amihai Mazar of the Hebrew University of Jerusalem in the ruins of the city, dating from about 900 BCE. The hives were found in orderly rows, three high, in a manner that could have accommodated around 100 hives, held more than 1 million bees and had a potential annual yield of 500 kilograms of honey and 70 kilograms of beeswax, according to Mazar, and are evidence that an advanced honey industry existed in ancient Israel 3,000 years ago. Ezra Marcus, an expert from the University of Haifa, said the finding was a glimpse of ancient beekeeping seen in texts and ancient art from the Near East. In ancient Greece, aspects of the lives of bees and beekeeping are discussed at length by Aristotle. Beekeeping was also documented by the Roman writers Virgil, Gaius Julius Hyginus, Varro, and Columella.

Study of Honey Bees

It was not until the 18th century that European natural philosophers undertook the scientific study of bee colonies and began to understand the complex and hidden world of bee biology. Preeminent among these scientific pioneers were Swammerdam, René Antoine Ferchault de Réaumur, Charles Bonnet, and the blind Swiss scientist Francois Huber. Swammerdam and Réaumur were among the first to use a microscope and dissection to understand the internal biology of honey bees. Réaumur was among the first to construct a glass walled observation hive to better observe activities within hives. He observed queens laying eggs in open cells, but still had no idea of how a queen was fertilized; nobody had ever witnessed the mating of a queen and drone and many theories held that queens were "self-fertile," while others believed that a vapor or "miasma" emanating from the drones fertilized queens without direct physical contact. Huber was the first to prove by observation and experiment that queens are physically inseminated by drones outside the confines of hives, usually a great distance away.

Following Réaumur's design, Huber built improved glass-walled observation hives and sectional hives which could be opened, like the leaves of a book, to inspect individual wax combs; this greatly improved the direct observation of activity within a hive. Although he became blind before he was twenty, Huber employed a secretary, Francois Burnens, to make daily observations, conduct careful experiments, and to keep accurate notes over a period of more than twenty years.

Huber confirmed that a hive consists of one queen who is the mother of all the female workers and male drones in the colony. He was also the first to confirm that mating with drones takes place outside of hives and that queens are inseminated by a number of successive matings with male drones, high in the air at a great distance from their hive. Together, he and Burnens dissected bees under the microscope and were among the first to describe the ovaries and spermatheca, or sperm store, of queens as well as the penis of male drones. Huber is universally regarded as "the father of modern bee-science" and his "Nouvelles Observations sur Les Abeilles (or "New Observations on Bees") revealed all the basic scientific truths for the biology and ecology of honeybees.

Invention of the Movable Comb Hive

Early forms of honey collecting entailed the destruction of the entire colony when the honey was harvested. The wild hive was crudely broken into, using smoke to suppress the bees, the honeycombs were torn out and smashed up — along with the eggs, larvae and honey they contained. The liquid honey from the destroyed brood nest was crudely strained through a sieve or basket. This was destructive and unhygienic, but for hunter-gatherer societies this did not matter, since the honey was generally consumed immediately and there were always more wild colonies to exploit. But in settled societies the destruction of the bee colony meant the loss of a valuable resource; this drawback made beekeeping both inefficient and something of a "stop and start" activity. There could be no continuity of production and no possibility of selective breeding, since each bee colony was destroyed at harvest time, along with its precious queen. During the medieval period abbeys and monasteries were centres of beekeeping, since beeswax was highly prized for candles and fermented honey was used to make alcoholic mead in areas of Europe where vines would not grow.

The 18th and 19th centuries saw successive stages of a revolution in beekeeping, which allowed the bees themselves to be preserved when taking the harvest.

Intermediate stages in the transition from the old beekeeping to the new were recorded for example by Thomas Wildman in 1768/1770, who described advances over the destructive old skep-based beekeeping so that the bees no longer had to be killed to harvest the honey. Wildman for example fixed a parallel array of wooden bars across the top of a straw hive or skep (with a separate straw top to be fixed on later) "so that there are in all seven bars of deal" [in a 10-inch diameter hive] "to which the bees fix their combs".

He also described using such hives in a multi-storey configuration, foreshadowing the modern use of supers: he described adding (at a proper time) successive straw hives below, and eventually removing the ones above when free of brood and filled with honey, so that the bees could be separately preserved at the harvest for a following season. Wildman also described a further development, using hives with "sliding frames" for the bees to build their comb, foreshadowing more modern uses of movable-comb hives. Wildman's book acknowledged the advances in knowledge of bees previously made by Swammerdam, Maraldi, and de Reaumur—he included a lengthy translation of Reaumur's account of the natural history of bees—and he also described the initiatives of others in designing hives for the preservation of bee-life when taking the harvest, citing in particular reports from Brittany dating from the 1750s, due to Comte de la Bourdonnaye.

The 19th Century saw this revolution in beekeeping practice completed through the perfection of the movable comb hive by Lorenzo Lorraine Langstroth, a descendant of Yorkshire farmers who emigrated to the United States. Langstroth was the first person to make practical use of Huber's earlier discovery that there was a specific spatial measurement between the wax combs, later called "the bee space", which bees would not block with wax, but kept as a free passage. Having determined this "bee space" (between 5 and 8 mm, or 1/4 to 3/8"), Langstroth then designed a series of wooden frames within a rectangular hive box, carefully maintaining the correct space between successive frames, and found that the bees would build parallel honeycombs in the box without bonding them to each other or to the hive walls.

This enables the beekeeper to slide any frame out of the hive for inspection, without harming the bees or the comb, protecting the eggs, larvae and pupae contained within the cells. It also meant that combs containing honey could be gently removed and the honey extracted without destroying the comb. The emptied honey combs could then be returned to the bees intact for refilling. Langstroth's classic book, The Hive and Honey-bee, published in 1853, described his rediscovery of the bee space and the development of his patent movable comb hive.

The invention and development of the movable-comb-hive fostered the growth of commercial honey production on a large scale in both Europe and the USA.

Evolution of Hive Designs

Langstroth's design for moveable comb hives was seized upon by apiarists and inventors on both sides of the Atlantic and a wide range of moveable comb hives were designed and perfected in England, France, Germany and the United States. Classic designs evolved in each country: Dadant hives and Langstroth hives are still dominant in the USA; in France the De-Layens trough-hive became popular and in the UK a British National Hive became standard as late as the 1930s although in Scotland the smaller Smith hive is still popular. In some Scandinavian countries and in Russia the traditional trough hive persisted until late in the 20th Century and is still kept in some areas. However, the Langstroth and Dadant designs remain ubiquitous in the USA and also in many parts of Europe, though Sweden, Denmark, Germany, France and Italy all have their own national hive designs. Regional variations of hive evolved to reflect the climate, floral productivity and the reproductive characteristics of the various subspecies of native honey bee in each bio-region.

The differences in hive dimensions are insignificant in comparison to the common factors in all these hives: they are all square or rectangular; they all use moveable wooden frames; they all consist of a floor, brood-box, honey-super, crown-board and roof. Hives have traditionally been constructed of cedar, pine, or cypress wood, but in recent years hives made from injection molded dense polystyrene have become increasingly important.

Hives also use queen excluders between the brood-box and honey supers to keep the queen from laying eggs in cells next to those containing honey intended for consumption. Also, with the advent in the 20th century of mite pests, hive floors are often replaced for part of (or the whole) year with a wire mesh and removable tray.

Pioneers of Practical and Commercial Beekeeping

The 19th Century produced an explosion of innovators and inventors who perfected the design and production of beehives, systems of management and husbandry, stock improvement by selective breeding, honey extraction and marketing. Preeminent among these innovators were:

Jan Dzierlon, was the father of modern apiology and apiculture. All modern beehives are descendants of his design.

L. L. Langstroth, Revered as the "father of American apiculture", no other individual has influenced modern beekeeping practice more than Lorenzo Lorraine

Langstroth. His classic book The Hive and Honey-bee was published in 1853. Moses Quinby, often termed 'the father of commercial beekeeping in the United States', author of Mysteries of Bee-Keeping Explained.

Amos Root, author of the A B C of Bee Culture which has been continuously revised and remains in print to this day. Root pioneered the manufacture of hives and the distribution of bee-packages in the United States.

A.J. Cook, author of The Bee-Keepers' Guide; or Manual of the Apiary, 1876.

Dr. C.C. Miller was one of the first entrepreneurs to actually make a living from apiculture. By 1878 he made beekeeping his sole business activity. His book, Fifty Years Among the Bees, remains a classic and his influence on bee management persists to this day.

Major Francesco De Hruschka was an Italian military officer who made one crucial invention that catalyzed the commercial honey industry. In 1865 he invented a simple machine for extracting honey from the comb by means of centrifugal force. His original idea was simply to support combs in a metal framework and then spin them around within a container to collect honey as it was thrown out by centrifugal force. This meant that honeycombs could be returned to a hive undamaged but empty — saving the bees a vast amount of work, time and materials. This single invention greatly improved the efficiency of honey harvesting and catalysed the modern honey industry.

In the UK practical beekeeping was led in the early 20th century by a few men, pre-eminently Brother Adam and his Buckfast bee and R.O.B. Manley, author of many titles, including 'Honey Production In The British Isles' and inventor of the Manley frame, still universally popular in the UK.

Other notable British pioneers include William Herrod-Hempsall and Gale.

Current practices of British beekeeping are relatively modern, due to the Acarapis woodi which destroyed almost all native bees in 1915.

Traditional Beekeeping

Fixed Frame Hives

There are considerable regional variations in the type of hive in which bees are kept. A hive is a set of rectangular wooden boxes filled with moveable wood or plastic frames, each of which holds a sheet of wax or plastic foundation. The bees build cells upon the sheets of foundation to create complete honeycombs. Foundation comes in two cell-sizes: worker foundation, which enables the bees to create small, hexagonal worker cells; and drone foundation, which allows the bees to build much larger drone cells, for the production of male bees.

The bottom box, or brood chamber, contains the queen and most of the bees; the upper boxes, or supers, contain just honey. Only the young nurse bees can produce wax flakes which they secrete from between their abdominal plates; they build honeycomb using the artificial wax foundation as a starting point, after which they may raise brood or deposit honey and pollen in the cells of the comb. These frames can be freely manipulated and honey supers with frames full of honey can be taken and extracted for their honey crop.

Modern Beekeeping

Movable Frame Hives

In the United States, the Langstroth hive is commonly used. The Langstroth was the first successful top-opened hive with movable frames, and other designs of hive have been based on it. Langstroth hive was however a descendant of Jan Dzierzon's Polish hive designs. In the United Kingdom, the most common type of hive is the British National Hive, which can hold Hoffman, British Standard or popular Manley frames, but it is not unusual to see some other sorts of hive (Smith, Commercial and WBC, rarely Langstroth). Straw skeps, bee gums, and unframed box hives are now unlawful in most US states, as the comb and brood cannot be inspected for diseases. However, straw skeps are still used for collecting swarms by hobbyists in the UK, before moving them into standard hives.

Top Bar Hives

A few hobby beekeepers are adopting various top bar hives of the type commonly found in Africa. Top bar hives were originally used as traditional beekeeping a method in both Greece and Viet Nam. These have no frames and the honey filled comb is not returned to the hive after extraction, as it is in the Langstroth hive. Because of this, the production of honey in a top bar hive is only about 20% that of a Langstroth hive, but the initial costs and equipment requirements are far lower. Top-bar hives also offer some advantages in interacting with the bees and the amount of weight that must be lifted is greatly reduced. Top Bar Hives are being widely used in developing countries in Africa and Asia as a result of the 'Bees For Development' program.

Protective Clothing

While knowledge of the bees is the first line of defense, most beekeepers also wear some protective clothing. Novice beekeepers usually wear gloves and a hooded suit or hat and veil. Experienced beekeepers sometimes elect not to use gloves because they inhibit delicate manipulations. The face and neck are the most important areas to protect, so most beekeepers will at least wear a veil.

It's worth noting that no amount of protective clothing will make the experience of a facefull of aggressive bees flying up from an opened hive pleasant for any beekeeper, and so it's rewarding to colonise kindly bees as soon as possible. Defensive bees are attracted to the breath, and a sting on the face can lead to much more pain and swelling than a sting elsewhere, while a sting on a bare hand can usually be quickly removed by fingernail scrape to reduce the amount of venom injected.

The protective clothing is generally light coloured (but not colourful) and of a smooth material. This provides the maximum differentiation from the colony's natural predators (bears, skunks, etc.) which tend to be dark-colored and furry.

The 'stings' retained in the fabric of the clothing will continue to pump out an alarm pheromone that actually attracts aggressive action and further stinging attacks. Washing suits regularly, and rinsing gloved hands in vinegar will minimise attraction.

Smoker

Smoke is the beekeeper's third line of defence. Most beekeepers use a "smoker" — a device designed to generate smoke from the incomplete combustion of various fuels. Smoke calms bees; it initiates a feeding response in anticipation of possible hive abandonment due to fire. Smoke also masks alarm pheromones released by guard bees or when bees are squashed in an inspection. The ensuing confusion creates an opportunity for the beekeeper to open the hive and work without triggering a defensive reaction. In addition, when a bee consumes honey the bee's abdomen distends, supposedly making it difficult to make the necessary flexes to sting, though this has not been tested scientifically.

Smoke is of questionable use with a swarm, because swarms do not have honey stores to feed on in response. Usually smoke is not needed, since swarms tend to be less defensive, as they have no stores to defend, and a fresh swarm will have fed well from the hive.

Many types of fuel can be used in a smoker as long as it is natural and not contaminated with harmful substances. These fuels include hessian, twine, burlap, pine needles, corrugated cardboard, and mostly rotten or punky wood. Indian beekeepers, especially in Kerala, often use coconut fibers as they are readily available, safe, and of negligible expense. Some beekeeping supply sources also sell commercial fuels like pulped paper and compressed cotton, or even aerosol cans of smoke. Other beekeepers use sumac as fuel because it ejects lots of smoke and doesn't have an odor.

Some bee keepers are using "liquid smoke" as a safer, more convenient, alternative. It is a water-based solution that is sprayed onto the bees from a plastic spray bottle.

Effects of Stings and of Protective Measures

Some beekeepers believe that the more stings a beekeeper receives, the less irritation each causes, and they consider it important for safety of the beekeeper to be stung a few times a season. Beekeepers have high levels of antibodies (mainly IgG) reacting to the major antigen of bee venom, phospholipase A2 (PLA). Antibodies correlate with the frequency of bee stings.

The entry of venom into the body from bee-stings may also be hindered and reduced by protective clothing which allows the wearer to remove stings and venom sacs simply with a tug on the clothing.

Natural Beekeeping

There is a current movement that eschews chemicals in beekeeping, and feels that Colony collapse disorder can be most effectively addressed by reversing trends that disrespect the needs of the bees themselves. Crop spraying, unnatural conditions in which bees are moved thousands of miles to pollinate commercial crops, artificial insemination of queens, and sugar water feeding are thought to all contribute to a general weakening of the constitution of the honeybee.

'Natural Beekeeping' by Ross Conrad, is one resource for learning more about this, as is the natural beekeeping network a global forum on the subject.

Passage from Natural Beekeeping: "Since the honey bee exerts such a powerfully beneficial influence on the natural world around us, it seems logical to assume that our own efforts to help the honey bee thrive can indirectly benefit all of nature."

The Backwards Beekeepers take many of their ideas from the writings of Charles Martin Simon, who urges a chemical-free and hands-off approach to beekeeping.

Urban or Backyard Beekeeping

Related to natural beekeeping, urban beekeeping is an attempt to revert to a less industrialized way of obtaining honey by utilizing small-scale colonies that pollinate urban gardens. Urban apiculture has undergone a renaissance in the 2000s. Paris, Berlin, London, Tokyo and Washington, D.C., are among beekeeping cities.

Until 2010, beekeeping was banned in New York City and punishable with a $2,000 fine. Urban beekeeping is commonly practiced in areas that have a pesticide ban. This includes Paris, as well as 156 municipalities in Canada and 3 of 10 Canadian provinces. Beekeeping was illegal in Vancouver until 2003, for example, but by 2010 there were bees on the roof of Vancouver City Hall.

Backyard beekeeping is not only a great hobby that rewards you with delicious, natural honey, but a fantastic way to help save the honey bees. Almost anyone can start beekeeping in his backyard or even in the flat roof. Not a lot of space is needed to place one or two hives. Not even, a garden. Honeybees can fly up to 5 miles in search of flowers. The only thing you must provide is a small source of water, such a bird bath or a shallow container filled with water. By keeping bees in urban areas, we are helping bees to not disappear and we are helping the gardens in the area. Gardens will be generously pollinated and this will improve the quantity and quality of flowers, fruits and vegetables.

Bee Colonies

Castes

A colony of bees consists of three castes of bee:

- a Queen bee, which is normally the only breeding female in the colony;
- a large number of female worker bees, typically 30,000–50,000 in number;
- a number of male drones, ranging from thousands in a strong hive in spring to very few during dearth or cold season.

The queen is the only sexually mature female in the hive and all of the female worker bees and male drones are her offspring. The queen may live for up to three years or more and may be capable of laying half a million eggs or more in her lifetime. At the peak of the breeding season, late spring to summer, a good queen may be capable of laying 3,000 eggs in one day, more than her own body weight. This would be exceptional however; a prolific queen might peak at 2,000 eggs a day, but a more average queen might lay just 1500 eggs per day. The queen is raised from a normal worker egg, but is fed a larger amount of royal jelly than a normal worker bee, resulting in a radically different growth and metamorphosis. The queen influences the colony by the production and dissemination of a variety of pheromones or 'queen substances'. One of these chemicals suppresses the development of ovaries in all the female worker bees in the hive and prevents them laying eggs.

Mating of Queens

The queen emerges from her cell after 15 days of development and she remains in the hive for 3–7 days before venturing out on a mating flight. Mating flight is otherwise known as 'nuptial flight'. Her first orientation flight may only last a few seconds, just enough to mark the position of the hive. Subsequent mating flights may last from 5 minutes to 30 minutes, and she may mate with a number of male drones on each flight. Over several matings, possibly a dozen or more, the queen will receive and store enough sperm from a succession of drones to fertilize hundreds of thousands of eggs. If she does not manage to leave the hive to mate — possibly due to bad weather or being trapped within part of the hive — she will remain infertile and become a 'drone layer', incapable of producing female worker bees, and the hive is doomed.

Mating takes place at some distance from the hive and often several hundred feet up in the air; it is thought that this separates the strongest drones from the weaker ones-ensuring that only the fastest and strongest drones get to pass on their genes.

Female Worker Bees

Almost all the bees in a hive are female worker bees. At the height of summer when activity in the hive is frantic and work goes on non-stop, the life of a worker bee may be as short as 6 weeks; in late autumn, when no brood is being raised and no nectar is being harvested, a young bee may live for 16 weeks, right through the winter. During its life a worker bee performs different work functions in the hive which are largely dictated by the age of the bee.

Period	*Work activity*
Days 1-3	Cleaning cells and incubation
Day 3-6	Feeding older larvae
Day 6-10	Feeding younger larvae
Day 8-16	Receiving honey and pollen from field bees
Day 12-18	Wax making and cell building
Day 14 onwards	Entrance guards; nectar and pollen foraging

Male Bees (Drones)

Drones are the largest bees in the hive (except for the queen), at almost twice the size of a worker bee. They do no work, do not forage for pollen or nectar and are only produced in order to mate with new queens and fertilize them on their mating flights. A bee colony will generally start to raise drones a few weeks before building queen cells in order to supersede a failing queen or in preparation for swarming. When queen raising for the season is over, the bees in colder climates will drive the drones out of the hive to die, biting and tearing at their legs and wings.

Differing Stages of Development

Stage of development	Queen	Worker	Drone
Egg	3 days	3 days	3 days
Larva	8 days	10 days	13 days
Pupa	4 days	8 days	8 days
Total	15 days	21 days	24 days

Structure of a Bee Colony

A domesticated bee colony is normally housed in a rectangular hive body, within which eight to ten parallel frames house the vertical plates of honeycomb which contain the eggs, larvae, pupae and food for the colony. If one were to cut a vertical cross-section through the hive from side to side, the brood nest would appear as a roughly ovoid ball spanning 5-8 frames of comb. The two outside combs at each side of the hive tend to be exclusively used for long-term storage of honey and pollen.

Within the central brood nest, a single frame of comb will typically have a central disk of eggs, larvae and sealed brood cells which may extend almost to the edges of the frame. Immediately above the brood patch an arch of pollen-filled cells extends from side to side, and above that again a broader arch of honey-filled cells extends to the frame tops. The pollen is protein-rich food for developing larvae, while honey is also food but largely energy rich rather than protein rich. The nurse bees which care for the developing brood secrete a special food called 'royal jelly' after feeding themselves on honey and pollen. The amount of royal jelly which is fed to a larva determines whether it will develop into a worker bee or a queen.

Apart from the honey stored within the central brood frames, the bees store surplus honey in combs above the brood nest. In modern hives the beekeeper places separate boxes, called 'supers', above the brood box, in which a series of shallower combs is provided for storage of honey. This enables the beekeeper to remove some of the supers in the late summer, and to extract the surplus honey harvest, without damaging the colony of bees and its brood nest below. If all the honey is 'stolen', including the amount of honey needed to survive winter, the beekeeper must replace these stores by feeding the bees sugar or corn syrup in autumn.

Annual Cycle of a Bee Colony

The development of a bee colony follows an annual cycle of growth which begins in spring with a rapid expansion of the brood nest, as soon as pollen is available for feeding larvae. Some production of brood may begin as early as January, even in a cold winter, but breeding accelerates towards a peak in May (in the northern hemisphere), producing an abundance of harvesting bees synchronised to the main 'nectar flow' in that region. Each race of bees times this build-up slightly differently, depending on how the flora of its original region blooms. Some regions of Europe have two nectar flows: one in late spring and another in late August. Other regions have only a single nectar flow. The skill of the beekeeper lies in predicting when the nectar flow will occur in his area and in trying to ensure that his colonies achieve a maximum population of harvesters at exactly the right time.

The key factor in this is the prevention, or skillful management of the swarming impulse. If a colony swarms unexpectedly and the beekeeper does not manage to capture the resulting swarm, he is likely to harvest significantly less honey from that hive, since he will have lost half his worker bees at a single stroke. If, however, he can use the swarming impulse to breed a new queen but keep all the bees in the colony together, he will maximize his chances of a good harvest. It takes many years of learning and experience to be able to manage all these aspects successfully, though owing to variable circumstances many beginners will often achieve a good honey harvest.

Formation of New Colonies

Colony Reproduction; Swarming and Supersedure

All colonies are totally dependent on their queen, who is the only egg-layer. However, even the best queens live only a few years and one or two years longevity is the norm. She can choose whether or not to fertilize an egg as she lays it; if she does so, it develops into a female worker bee; if she lays an unfertilized egg it becomes a male drone. She decides which type of egg to lay depending on the size of the open brood cell which she encounters on the comb; in a small worker cell she lays a fertilized egg; if she finds a much larger drone cell she lays an unfertilized drone egg.

All the time that the queen is fertile and laying eggs she produces a variety of pheromones which control the behaviour of the bees in the hive; these are commonly called 'queen substance' but in reality there are various different pheromones with different functions. As the queen ages she begins to run out of stored sperm and her pheromones begin to fail. At some point, inevitably, the queen begins to falter and the bees will decide to replace her by creating a new queen from one of her worker eggs. They may do this because she has been damaged (lost a leg or an antenna), because she has run out of sperm and cannot lay fertilized eggs (has become a 'drone laying queen') or because her pheromones have dwindled to a point where they cannot control all the bees in the hive anymore.

At this juncture the bees will produce one or more queen cells by modifying existing worker cells which contain a normal female egg.

However, there are two distinct behaviours which the bees pursue:

1. Supersedure: queen replacement within one hive without swarming
2. Swarm cell production: the division of the hive into two colonies by swarming.

Different sub-species of Apis mellifera exhibit differing swarming characteristics which reflect their evolution in different ecotopes of the European continent. In general the more northerly black races are said to swarm less and supersede more, whereas the more southerly yellow and grey varieties are said to swarm more frequently. The truth is complicated because of the prevalence of cross-breeding and hybridization of the sub species and opinions differ.

Supersedure is highly valued as a behavioural trait by beekeepers because a hive that supersedes its old queen does not swarm and so no stock is lost; it merely creates a new queen and allows the old one to fade away, or alternatively she is killed when the new queen emerges. When superseding a queen the bees will produce just one or two queen cells, characteristically in the center of the face of a broodcomb.

In swarming, by contrast, a great many queen cells are created — typically a dozen or more — and these are located around the edges of a broodcomb, most often at the sides and the bottom.

Once either process has begun, the old queen will normally leave the hive with the hatching of the first queen cells. When she leaves the hive the old queen is accompanied by a large number of bees, predominantly young bees (wax-secreters), who will form the basis of the new hive. Scouts are sent out from the swarm to find suitable hollow trees or rock crevices and as soon as one is found the entire swarm moves in, building new wax brood combs within a matter of hours using the honey stores which the young bees have filled themselves with before leaving the old hive. Only young bees can secrete wax from special abdominal segments and this is why there tend to be more young bees than old in swarms. Often a number of virgin queens accompany the first swarm (the 'prime swarm'), and the old queen is replaced as soon as a daughter queen is mated and laying. Otherwise, she will be quickly superseded in their new home.

Factors that Trigger Swarming

It is generally accepted that a colony of bees will not swarm until they have completed all of their brood combs, i.e. filled all available space with eggs, larvae and brood. This generally occurs in late Spring at a time when the other areas of the hive are rapidly filling with honey stores. So one key trigger of the swarming instinct is when the queen has no more room to lay eggs and the hive population is becoming very congested. Under these conditions a prime swarm may issue with the queen, resulting in a halving of the population within the hive and leaving the old colony with a large amount of hatching bees. The queen who leaves finds herself in a new hive with no eggs, no larvae but lots of energetic young bees who create a new set of brood combs from scratch in a very short time.

Another important factor in swarming is the age of the queen. Those under a year in age are unlikely to swarm unless they are extremely crowded, while older queens have swarming predisposition.

Beekeepers monitor their colonies carefully in spring and watch for the appearance of queen cells, which are a dramatic signal that the colony is determined to swarm.

When a colony has decided to swarm, queen cells are produced in numbers varying to a dozen or more. When the first of these queen cells is sealed, after 8 days of larval feeding, a virgin queen will pupate and be due to emerge seven days after sealing. Before leaving, the worker bees fill their stomachs with honey in preparation for the creation of new honeycombs in a new home. This cargo of honey also makes swarming bees less inclined to sting and a newly issued swarm is noticeably gentle for up to 24 hours — often capable of being handled without gloves or veil by a beekeeper.

This swarm is looking for shelter. A beekeeper may capture it and introduce it into a new hive, helping to meet this need. Otherwise, it will return to a feral state, in which case it will find shelter in a hollow tree, an excavation, an abandoned chimney, or even behind shutters.

Back at the original hive, the first virgin queen to emerge from her cell will immediately seek to kill all her rival queens who are still waiting to emerge from their cells. However, usually the bees deliberately prevent her from doing this, in which case, she too will lead a second swarm from the hive. Successive swarms are called 'after-swarms' or 'casts' and can be very small, often with just a thousand or so bees, as opposed to a prime swarm which may contain as many as ten to twenty thousand bees.

Small after-swarms have less chance of survival and, depleting the original hive, may threaten its survival as well. When a hive has swarmed despite the beekeeper's preventative efforts, a good management practice is to give the depleted hive a couple frames of open brood with eggs. This helps replenish the hive more quickly, and gives a second opportunity to raise a queen, if there is a mating failure.

Each race or sub-species of honeybee has its own swarming characteristics. Italian bees are very prolific and inclined to swarm; Northern European black bees have a strong tendency to supersede their old queen, without swarming. These differences are the result of differing evolutionary pressures in the regions where each sub-species evolved.

Artificial Swarming

When a colony accidentally loses its queen, it is said to be 'queenless'. The workers realize that the queen is absent after as little as an hour, as her pheromones fade in the hive. The colony cannot survive without a fertile queen laying eggs to renew the population. So the workers select cells containing eggs aged less than three days and enlarge these cells dramatically to form 'emergency queen cells'. These appear similar to large peanut-like structure about an inch long, which hangs from the center or side of the brood combs.

The developing larva in a queen cell is fed differently from an ordinary worker-bee, receiving in addition to the normal honey and pollen a great deal of royal jelly, a special food secreted by young 'nurse bees' from the hypopharyngeal gland. This special food dramatically alters the growth and development of the larva so that, after metamorphosis and pupation, it emerges from the cell as a queen bee. The queen is the only bee in a colony which has fully developed ovaries and she secretes a pheromone which suppresses the normal development of ovaries in all her worker-daughters.

Beekeepers use the ability of the bees to produce new queens in order to increase their colonies, a procedure called splitting a colony. In order to do this, they remove several brood combs from a healthy hive, taking care that the old queen is left behind. These combs must contain eggs or larvae less than three days old which will be covered by young 'nurse bees' which care for the brood and keep it warm. These brood combs and attendant nurse bees are then placed into a small 'nucleus hive' along with other combs containing honey and pollen. As soon as the nurse bees find themselves in this new hive and realise that they have no queen they set about constructing emergency queen cells using the eggs or larvae which they have in the combs with them.

Bee Keeping (Apiculture) in India

Bee keeping is an agro based enterprise, which farmers can take up for additional income generation.

Honey bees convert nectar of flowers into honey and store them in the combs of the hive.

Collection of honey from the forests has been in existence for a long time. The growing market potential for honey and its products has resulted in bee keeping emerging as a viable enterprise. Honey and wax are the two economically important products of bee keeping.

Honey and beekeeping have a long history in India. Honey was the first sweet food tasted by the ancient Indian inhabiting rock shelters and forests. He hunted bee hives for this gift of god. India has some of the oldest records of beekeeping in the form of paintings by prehistoric man in the rock shelters. With the development of civilization, honey acquired an unique status in the lives of the ancient Indians. They regarded honey as a magical substance that controlled the fertility of women, cattle, as also their lands and crops. The recent past has witnessed a revival of the industry in the rich forest regions along the sub-Himalayan mountain ranges and the Western Ghats, where it has been practiced in its simplest form.

In India beekeeping has been mainly forest based. Several natural plant species provide nectar and pollen to honey bees. Thus, the raw material for production of honey is available free from nature. Bee hives neither demand additional land space nor do they compete with agriculture or animal husbandry for any input. The beekeeper needs only to spare a few hours in a week to look after his bee colonies. Beekeeping is therefore ideally suited to him as a part-time occupation. Beekeeping constitutes a resource of sustainable income generation to the rural and tribal farmers. It provides them valuable nutrition in the form of honey, protein rich pollen and brood. Bee products also constitute important ingredients of folk and traditional medicine.

The establishment of Khadi and Village Industries Commission to revitalize the traditional village industries, hastened the development of beekeeping. During the 1980s, an estimated one million bee hives had been functioning under various schemes of the Khadi and Village Industries Commission. Production of apiary honey in the country reached 10,000 tons, valued at about Rs. 300 million.

Side by side with the development of apiculture using the indigenous bee, Apis cerana, apiculture using the European bee, Apis mellifera, gained popularity in Jammu & Kashmir, Punjab, Himachal Pradesh, Haryana, Uttar Pradesh, Bihar and West Bengal. Wild honey bee colonies of the giant honey bee and the oriental hive bee have also been exploited for collection of honey. Tribal populations and forest dwellers in several parts of India have honey collection from wild honey bee nests as their traditional profession. The methods of collection of honey and beeswax from these nests have changed only slightly over the millennia. The major regions for production of this honey are the forests and farms along the sub-Himalayan tracts and adjacent foothills, tropical forest and cultivated vegetation in Rajasthan, Uttar Pradesh, Madhya Pradesh, Maharashtra and Eastern Ghats in Orissa and Andhra Pradesh.

Resources and Potential

The raw materials for the beekeeping industry are mainly pollen and nectar that come from flowering plants. Both the natural and cultivated vegetation in India constitute an immense potential for development of beekeeping. About 500 flowering plant species, both wild and cultivated, are useful as major or minor sources of nectar and pollen. There are at least four species of true honey bees and three species of the stingless bees. Several sub-species and races of these are known to exist. In recent years the exotic honey bee has been introduced. Together these represent a wide variety of bee fauna that can be utilized for the development of honey industry in the country. There are several types of indigenous and traditional hives including logs, clay pots, wall niches, baskets and boxes of different sizes and shapes. In modern beekeeping, the combs are built on wooden frames that are moveable. This facilitates inspection and management of bee colonies. Three types of moveable frame hive are in common use : the Newton type along with its standardized version ISI Type A, the Jeolikote Villager, and its counterpart ISI Type B, and the Langstroth type. Besides the hives, the beekeepers need equipment and implements like the hive stand, nucleus box and smoker. The industry also needs equipment and machinery for handling and processing of honey, beeswax, for manufacture of comb foundation sheets, and for other operations.

India has a potential to keep about 120 million bee colonies, that can provide self-employment to over 6 million rural and tribal families. In terms of production, these bee colonies can produce over 1.2 million tons of honey and about 15,000 tons of beeswax. Organized collection of forest honey and beeswax using improved methods can result in an additional production of at least 120,000 tons of honey and 10,000 tons of beeswax. This can generate income to about 5 million tribal families.

Apiculture Technology

Production of honey has been the major aim of the industry. Modern beekeeping also includes production of beeswax, bee collected pollen, bee venom, royal jelly, propolis, as also of package bees, queen bees and nucleus colonies. All these are possible only with a proper management of bees, utilizing the local plant resources and adapting to the local climatic conditions. Modern beekeeping makes heavy use of beekeeping equipment and honey processing plant. This results in high efficiency and also ensures the quality of the processed honey.

Seasonal management of bee colonies varies in different parts of the country although the basic management methods are the same. Flow management, dearth management, provision of feeding, and control and cure of bee disorders, bee diseases, pests and enemies, are some of the routine measures to keep bee colonies healthy and strong. There are special management techniques like queen rearing, migration for honey production or for colony multiplication, which the beekeeper takes up after he gains sufficient knowledge and experience in handling bee colonies.

Market

About 10,000 tons of forest honey are produced annually. Apiary honey produced under the KVI sector is estimated to be a little less than 10,000 tons in 1990-91. Over 95 per cent of this was from the A. cerana colonies, the rest being from the European bee colonies. Forest honey, mostly from rock bee hives, is usually collected by tribals in forests and is procured by forest or tribal corporations as a minor forest produce. Quite a large quantity is also collected by groups or individuals on their own. Forest honey is usually thin, contains large quantity of pollen, bee juices and parts, wax and soil particles. The honey collector gets between Rs. 10 and Rs. 25 per kilogram of the forest honey. Forest honeys are mostly multifloral.

Apiary honey is produced in bee hives and is harvested by extraction in honey extractors. Other types of beekeeping equipment like queen excluder, smoker, hive tool, pollen trap, honey processing plant are also used. Beekeepers sell the honey to the co-operative society, if one exists in the area. In many parts of India, the beekeeper gets a much higher price if he sells it directly to the consumer. Apiary honeys are usually multifloral when marketed by state-level marketing organizations, because honeys from different sources are mixed while pooling, storage and processing. Several unifloral honeys are available in markets restricted to small areas within the state where it is produced. Rubber plant contributed to over 60 per cent of the total apiary honey production during 1990-91. Besides this, jamun, hirda, beheda, arjun, neem, litchi, palmyrah palm, eucalyptus, lagerstroemias, tamarind, cashew tree, scheffleras, tun, karanj, false acacia, wild shrubs like shain, crops of different varieties of mustards, sesame, niger, sunflower, berseem clover, khesari, coriander, orchard trees including different types of citrus, apple, puddum, cherry and other temperate fruit trees, coconut tree and coffee plantations are some important sources that provide unifloral honey.

Much of the forest honey is sold to the pharmaceutical, confectionery and food industries, where it is processed and used in different formulations. Apiary honey is usually processed at the producers level. This consists mainly of heating the honey and filtering. A few beekeepers or honey producers co-operative societies have better processing facilities that involve killing of honey fermenting yeasts. About 50 per cent of the apiary honey under the KVI sector is graded and marketed under AGMARK specifications. In 1985 the consumption of honey was estimated to be about 8.4 g per capita, while in other countries this was 200 g. Presently this would be about 2.5 g. Honey has so far been consumed mainly as a medicine and for religious purposes. A small quantity has been used in kitchen as an ingredient of pickles, jams and preserves. With the increasing production in recent years, there is an increasing trend to use honey in food. This is obviously the case with the affluent segments of the population. Forest honey is used in pharmaceutical, food, confectionery, bakery and cosmetic industries.

One often finds a good demand for local honeys like honeys from Mahabaleshwar. People in Maharashtra have a strong liking for jamun, hirda or gela honeys which have acquired special individual medicinal significance. Similarly, kartiki honey in Kumaon, Uttar Pradesh is locally much favoured. Some honeys have an essentially non-local market. Rubber tree honey can only be sold in non-local markets. Coorg honey with its characteristic flavour was well-known during 1950s and 1960s. Shain or sulah honey from Kashmir has been very popular. Presently litchi honey from Bihar and other northern states is in great demand. The price structure is regulated by the market forces of supply and demand. Beekeepers in well-known hill stations and other places of tourist attraction take advantage of the popularity of honey and can market their produce at remunerative prices.

Indian honey has a good export market. With the use of modern collection, storage, beekeeping equipment, honey processing plants and bottling technologies the potential export market can be tapped.

Advantages of beekeeping as an income generation activity:

- Bee keeping requires less time, money and infrastructure investments
- Honey and beeswax can be produced from an area of little agricultural value
- The Honey bee does not compete for resources with any other agricultural enterprise.
- Beekeeping has positive ecological consequences. Bees play an important role in the pollination of many flowering plants, thus increasing the yield of certain crops such as sunflower and various fruits.
- Honey is a delicious and highly nutritious food. By the traditional method of honey hunting many wild colonies of bees are destroyed. This can be prevented by raising bees in boxes and producing honey at home.
- Beekeeping can be initiated by individuals or groups
- The market potential for honey and wax is high.

Production Process

Equipment Requirements for Bee Keeping:

- Hive: It is a simple long box covered with a number of slats on top. The rough measurements of the box should be around 100 cm of length, 45 cm of width and 25 cm in height. The box should be 2 cm thick and the hive must be glued and screwed together with entrance holes of 1 cm wide. The slats (top bars) must be as long as the hive is wide in order to fit across and the thickness of about 1.5 cm is sufficient to support a heavy honey comb. The width of 3.3 cm needs to be given to give the bees the natural spacing they need to easily build one comb to each separate top bar.

- Smoker : It is the second important piece of equipment. This can be made from a small tin. We use the smoker to protect ourselves from bee stings and to control the bees.
- Cloth: to protect our eyes and nose from stings at the time of work near the apiary.
- Knife: It is used to loosen the top bars and to cut of the honey bars.
- Feather: To sweep the bees from the comb.
- Queen Excluder
- Match box.

Species of Honey Bees

There are four species of honeybees in India. They are:

- Rock bee (Apis dorsata): They are good honey gathers with an average yield of 50-80 kg per colony.
- Little bee (Apis florea): They are poor honey yielders and yield about 200-900 g of honey per colony.
- Indian bee (Apis cerana indica): They yield an average honey yield of 6-8 kg per colony per year.
- European bee [Italian bee] (Apis mellifera): The average production per colony is 25-40 kg.

Stingless bee (Trigona iridipennis): In addition to the above, another species is also present in Kerala known as stingless bees. They are not truly stingless, but sting is poorly developed. They are efficient pollinators. They yield 300-400 g of honey per year.

Establishment of hives:

- The apiary must be located in well-drained open area, preferably near orchards, with profuse source of nectar, pollen and water.
- Protection from sunlight is important in order to maintain an optimum temperature in the hive.
- Ant wells are fixed around the hive stand. The colonies must be directed towards east, with slight changes in the directions of the bee box as a protection from rain and sun.

- Keep the colonies away from the reach of cattle, other animal, busy roads and streetlights.

Establishing a bee colony:

- To establish a bee colony, bees can be obtained by transferring a wild nesting colony to a hive or attract a passing swarm of bees to occupy it.
- Before putting a swarm or even a colony in a prepared hive, it would be beneficial to make the hive smell familiar by rubbing old brown comb pieces or some bee wax. If possible, the Queen bee can be captured from a natural swarm and placed under a hive to attract the other bees.
- Feed the hived swarm for a few weeks by diluting a half cup of white sugar in half a cup of hot water as this will also help in building the comb along with the bars rapidly.
- Avoid over crowding.

Management of colonies:

- Inspect the beehives at least once in a week during the honey-flow seasons preferably during the morning hours.
- Clean the hive in the following sequence, the roof, super/supers, brood chambers and floorboard.
- Observe the colonies regularly for the presence of healthy queen, brood development, storage of honey and pollen, presence of queen cells, bee strength and growth of drones.
- Look for the infestation by any of the following bee enemies.
- *Wax moth (*Galleria mellonella*): Remove all the larvae and silken webbings from the combs, corners and crevices of bee box.*
- *Wax beetles (*Platybolium sp.*): Collect and destroy the adult beetles.*
- Mites: Clean the frame and floorboard with cotton swabs moistened with freshly made potassium permanganate solution. Repeat until no mites are seen on the floorboard.
- Management during lean season
- Remove the supers and arrange the available healthy broods compactly in the brood chamber.
- Provide division board, if necessary.
- Destroy queen cells and drone cells, if noted.
- Provide sugar syrup (1:1) @ 200 g sugar per colony per week for Indian bees.
- Feed all the colonies in the apiary at the same time to avoid robbing.
- Management during honey flow season
- Keep the colony in sufficient strength before honey-flow season.

- Provide maximum space between the first super and the brood chamber and not above the first super.
- Place queen excluder sheets in between brood and super chamber to confine the queen to brood chamber.
- Examine the colony once in a week and frames full of honey should be removed to the sides of the super. The frames, which are three-fourth filled with honey or pollen and one-fourth with sealed brood should be taken out of brood chamber and in its place empty combs or frames with foundation is added.
- The combs, which are completely sealed, or two-third capped may be taken out for extraction of honey and returned to supers after honey extraction.

Harvesting of honey:

- Harvest the honey by smoking the bees off the parts which needs to be harvested and cut the combs carefully.
- Harvests are normally possible during and shortly after the two main flowering seasons, namely October/November and February-June.
- A ripe comb is light in colour and filled with honey. More than half of the honey cells on both the sides are sealed with wax.

Rabbit

Rabbits are small mammals in the family Leporidae of the order Lagomorpha, found in several parts of the world. There are seven different genera in the family classified as rabbits, including the European rabbit (Oryctolagus cuniculus), Cottontail rabbit (genus Sylvilagus; 13 species), and the Amami rabbit (Pentalagus furnessi, endangered species on Amami Ôshima, Japan). There are many other species of rabbit, and these, along with pikas and hares, make up the order Lagomorpha.

Location and Habitat

The rabbit lives in many areas around the world. Rabbits live in groups, and the best known species, the European rabbit lives in underground burrows, or rabbit holes. A group of burrows is called a warren. Meadows, woods, forests, thickets, and grass lands are areas in which rabbits live. They also inhabit deserts and wetlands. More than half the world's rabbit population resides in North America. They also live in Europe, India, Sumatra, Japan, and parts of Africa. The European rabbit has been introduced to many places around the world.

Characteristics and Anatomy

The rabbit's long ears, which can be more than 10 cm (4 in) long, are probably an adaptation for detecting predators. They have large, powerful hind legs. Each foot has five toes, with one greatly reduced in size. They are digitigrade animals;

they move around on the tips of their toes. Wild rabbits do not differ much in their body proportions or stance, with full, egg-shaped bodies. Their size can range anywhere from 20 cm (8 in) in length and 0.4 kg in weight to 50 cm (20 in) and more than 2 kg. The fur is most commonly long and soft, with colors such as shades of brown, gray, and buff. The tail is a little plume of brownish fur (white on top for cottontails).

Cecal Pellets

Rabbits are hindgut digesters. This means that most of their digestion takes place in their large intestine and cecum. In rabbits, the cecum is about 10 times bigger than the stomach, and it, along with the large intestine, makes up roughly 40% of the rabbit's digestive tract. Cecotropes, sometimes called "night feces", come from the cecum and are high in minerals, vitamins and proteins that are necessary to the rabbit's health. Rabbits eat these to meet their nutritional requirements. This process allows rabbits to extract the necessary nutrients from their food.

Rabbits have two sets of incisor teeth, one behind the other. This way they can be distinguished from rodents, with which they are often mistaken.

Natural Behaviour

Rabbits, being prey animals, tend to be exploratory in new spaces and if confronted by a potential threat, tend to freeze and observe. Rabbits have a remarkably wide field of vision, and a good deal of it is devoted to overhead scanning. Both indoors and outdoors, rabbits will scan for overhead threats. They survive by burrowing, hopping away from danger and delivering powerful kicks with their hind legs. Their teeth are strong to allow them to eat and get out of struggle.

Reproduction

Rabbits have a very fast reproductive rate. The breeding season for most rabbits lasts 9 months, from February to October. Normal gestation is about 30 days. The average size of the litter varies but is usually between 4 and 12 babies, with larger breeds having larger litters. A kit (baby rabbit) can be weaned at about 4 to 5 weeks of age. This means in one season a single female rabbit can produce as many as 800 children, grandchildren, and great-grandchildren. A doe is ready to breed at about 6 months of age, and a buck at about 7 months. Courtship and mating is very brief, lasting only 30 to 40 seconds. Courtship behaviour involves licking, sniffing, and following the doe. Spraying urine is also a common sexual behaviour. Female rabbits are reflex ovulators.

Ovulation begins 10 hours after mating. After mating, the female will make a nest or borough, and line the nest with fur from the dewlap, flanks, and belly. This behaviour also exposes the nipples enabling her to better nurse the kits. Kits are altricial, which means they're born blind, naked, and helpless. Passive immunity (immunity acquired by transfer of antibodies or sensitized lymphocytes from another animal) is acquired by kits prior to birth via placental transfer. At about 10 to 11 days after birth, the baby rabbit's eyes will open, and they will start eating on their own at around 14 days old.

Although born naked, a soft baby coat of hair forms within a few days. At the age of 5 to 6 weeks, the soft baby coat is replaced with a pre-adult coat. At about 6 to 8 months of age, this intermediate coat is replaced by the final adult coat, which is shed twice a year thereafter. Due to the nutritious nature of rabbit milk, kits only need to be nursed for a few minutes once or twice a day.

Diet and Eating Habits

Rabbits are herbivores who feed by grazing on grass, forbs, and leafy weeds. In consequence, their diet contains large amounts of cellulose, which is hard to digest. Rabbits solve this problem by passing two distinct types of feces: hard droppings and soft black viscous pellets, the latter of which are immediately eaten. Rabbits reingest their own droppings (rather than chewing the cud as do cows and many other herbivores) to digest their food further and extract sufficient nutrients.

Rabbits graze heavily and rapidly for roughly the first half hour of a grazing period (usually in the late afternoon), followed by about half an hour of more selective feeding. In this time, the rabbit will also excrete many hard fecal pellets, being waste pellets that will not be reingested. If the environment is relatively non-threatening, the rabbit will remain outdoors for many hours, grazing at intervals. While out of the burrow, the rabbit will occasionally reingest its soft, partially digested pellets; this is rarely observed, since the pellets are reingested as they are produced. Reingestion is most common within the burrow between 8 o'clock in the morning and 5 o'clock in the evening, being carried out intermittently within that period.

Hard pellets are made up of hay-like fragments of plant cuticle and stalk, being the final waste product after redigestion of soft pellets. These are only released outside the burrow and are not reingested. Soft pellets are usually produced several hours after grazing, after the hard pellets have all been excreted. They are made up of micro-organisms and undigested plant cell walls.

The chewed plant material collects in the large cecum, a secondary chamber between the large and small intestine containing large quantities of symbiotic bacteria that help with the digestion of cellulose and also produce certain B vitamins. The pellets are about 56% bacteria by dry weight, largely accounting for the pellets being 24.4% protein on average. These pellets remain intact for up to six hours in the stomach; the bacteria within continue to digest the plant carbohydrates. The soft feces form here and contain up to five times the vitamins of hard feces. After being excreted, they are eaten whole by the rabbit and redigested in a special part of the stomach. This double-digestion process enables rabbits to use nutrients that they may have missed during the first passage through the gut, and thus ensures that maximum nutrition is derived from the food they eat. This process serves the same purpose within the rabbit as rumination does in cattle and sheep. Rabbits are incapable of vomiting due to the physiology of their digestive system.

Differences from Hares

Rabbits are clearly distinguished from hares in that rabbits are altricial, having young that are born blind and hairless. In contrast, hares are generally born with hair and are able to see (precocial). All rabbits except the cottontail rabbit live underground in burrows or warrens, while hares live in simple nests above the ground (as does the cottontail rabbit), and usually do not live in groups. Hares are generally larger than rabbits, with longer ears, and have black markings on their fur. Hares have not been domesticated, while European rabbits are often kept as house pets. In gardens, they are typically kept in hutches — small, wooden, house-like boxes — that protect the rabbits from the environment and predators.

As Pets

Pet rabbits kept indoors are referred to as house rabbits. House rabbits typically have an indoor pen or cage and a rabbit-safe place to run and exercise, such as an exercise pen, living room or family room. Rabbits can be trained to use a litter box and some can learn to come when called. Domestic rabbits that do not live indoors can also often serve as companions for their owners, typically living in an easily accessible hutch outside the home. Some pet rabbits live in outside hutches during the day for the benefit of fresh air and natural daylight and are brought inside at night.

Whether indoor or outdoor, pet rabbits' pens are often equipped with enrichment activities such as shelves, tunnels, balls, and other toys. Pet rabbits are often provided additional space in which to get exercise, simulating the open space a rabbit would traverse in the wild. Exercise pens or lawn pens are often used to provide a safe place for rabbits to run.

A pet rabbit's diet typically consists of unlimited Timothy hay, a small amount of pellets, and a small portion of fresh vegetables. Rabbits are social animals. Rabbits as pets can find their companionship with a variety of creatures, including humans, other rabbits, guinea pigs, and sometimes even cats and dogs. Animal welfare organisations such as the House Rabbit Society recommend that rabbits do not make good pets for small children because children generally do not know how to stay quiet, calm, and gentle around rabbits. As prey animals, rabbits are alert, timid creatures that startle easily. They have fragile bones, especially in their backs, that require support on the belly and bottom when picked up. Children 7 years old and older usually have the maturity required to care for a rabbit.

As Food and Clothing

Leporids such as European rabbits and hares are a food meat in Europe, South America, North America, some parts of the Middle East, and China, among other places. Rabbit is still commonly sold in UK butchers and markets, although not frequently in supermarkets. At farmers markets and the famous Borough Market in London, rabbits will be displayed dead and hanging unbutchered in the traditional style next to braces of pheasant and other small game. Rabbit meat was once commonly sold in Sydney, Australia, the sellers of which giving the name to the rugby league team the South Sydney Rabbitohs, but quickly became unpopular after the disease myxomatosis was introduced in an attempt to wipe out the feral rabbit population.

When used for food, rabbits are both hunted and bred for meat. Snares or guns along with dogs are usually employed when catching wild rabbits for food. In many regions, rabbits are also bred for meat, a practice called cuniculture. Rabbits can then be killed by hitting the back of their heads, a practice from which the term rabbit punch is derived. Rabbit meat is a source of high quality protein. It can be used in most ways chicken meat is used. In fact, well-known chef Mark Bittman says that domesticated rabbit tastes like chicken because both are blank palettes upon which any desired flavours can be layered. Rabbit meat is leaner than beef, pork, and chicken meat. Rabbit products are generally labelled in three ways, the first being Fryer. This is a young rabbit between 4.5 and 5 pounds and up to 9 weeks in age. This type of meat is tender and fine grained. The next product is a Roaster; they are usually over 5 pounds and up to 8 months in age. The flesh is firm and coarse grained and less tender than a fryer. Then there are giblets which include the liver and heart. One of the most common types of rabbit to be bred for meat is New Zealand white rabbit.

There are several health issues associated with the use of rabbits for meat, one of which is Tularemia or Rabbit Fever. Another is so-called rabbit starvation, due most likely to deficiency of essential fatty acids in rabbit meat. Rabbits are a common food item of large pythons, such as Burmese pythons and reticulated pythons, both in the wild and in captivity.

Rabbit pelts are sometimes used for clothing and accessories, such as scarves or hats. Angora rabbits are bread for their long, fine hair, which can be sheared and harvested like sheep wool. Rabbits are very good producers of manure; additionally, their urine, being high in nitrogen, makes lemon trees very productive. Their milk may also be of great medicinal or nutritional benefit due to its high protein content.

Environmental Problems

Rabbits have been a source of environmental problems when introduced into the wild by humans. As a result of their appetites, and the rate at which they breed, wild rabbit depredation can be problematic for agriculture. Gassing, barriers (fences), shooting, snaring, and ferreting have been used to control rabbit populations, but the most effective measures are diseases such as myxomatosis (myxo or mixi, colloquially) and calicivirus. In Europe, where rabbits are farmed on a large scale, they are protected against myxomatosis and calicivirus with a genetically modified virus.

The virus was developed in Spain, and is beneficial to rabbit farmers. If it were to make its way into wild populations in areas such as Australia, it could create a population boom, as those diseases are the most serious threats to rabbit survival. Rabbits in Australia are considered to be such a pest that land owners are legally obliged to control them.

In Culture and Literature

Rabbits are often used as a symbol of fertility or rebirth, and have long been associated with spring and Easter as the Easter Bunny. The species' role as a prey animal also lends itself as a symbol of innocence, another Easter connotation.

Additionally, rabbits are often used as symbols of playful sexuality, which also relates to the human perception of innocence, as well as its reputation as a prolific breeder.

Folklore and Mythology

The rabbit often appears in folklore as the trickster archetype, as he uses his cunning to outwit his enemies:

- In Aztec mythology, a pantheon of four hundred rabbit gods known as Centzon Totochtin, led by Ometotchtli or Two Rabbit, represented fertility, parties, and drunkenness.
- *In Central Africa "Kalulu" the rabbit is widely known as a tricky character, getting the better of bargains.*
- In Chinese literature, rabbits accompany Chang'e on the Moon. Also associated with the Chinese New Year (or Lunar New Year), rabbits are also one of the twelve celestial animals in the Chinese Zodiac for the Chinese calendar. It is interesting to note that the Vietnamese lunar new year replaced the rabbit with a cat in their calendar, as rabbits did not inhabit Vietnam.
- *In the folklore of the United States, a rabbit's foot is frequently carried as an amulet, and is often used on keychains, where it is thought to bring luck. The practice derives from the system of African-American folk magic called* hoodoo.
- In Japanese tradition, rabbits live on the Moon where they make mochi, the popular snack of mashed sticky rice. This comes from interpreting the pattern of dark patches on the moon as a rabbit standing on tiptoes on the left pounding on an usu, a Japanese mortar.
- In Jewish folklore, rabbits are associated with cowardice, a usage still current in contemporary Israeli spoken Hebrew (similar to English colloquial use of "chicken" to denote cowardice).

- A Korean myth similar to the Japanese counterpart presents rabbits living on the moon making rice cakes (Tteok in Korean).
- In Native American Ojibwe mythology, Nanabozho, or Great Rabbit, is an important deity related to the creation of the world.
- A Vietnamese mythological story portrays the rabbit of innocence and youthfulness. The Gods of the myth are shown to be hunting and killing rabbits to show off their power.
- "Taushan Tepe" (Rabbit Hill) was the Turkish name of Kabile, Bulgaria.

On the Isle of Portland in Dorset, UK, the rabbit is said to be unlucky and speaking its name can cause upset with older residents. This is thought to date back to early times in the quarrying industry, where piles of extracted stone (not fit for sale) were built into tall rough walls (to save space) directly behind the working quarry face; the rabbit's natural tendency to burrow would weaken these "walls" and cause collapse, often resulting in injuries or even death. The name rabbit is often substituted with words such as "long ears" or "underground mutton", so as not to have to say the actual word and bring bad luck to oneself. It is said that a public house (on the island) can be cleared of people by calling out the word rabbit and while this was very true in the past, it has gradually become more fable than fact over the past 50 years.

Other Fictional Rabbits

The rabbit as trickster appears in American popular culture; for example the Br'er Rabbit character from African-American folktales and Disney animation; and the Warner Bros. cartoon character Bugs Bunny.

Anthropomorphized rabbits have appeared in a host of works of film, literature, and technology, notably the White Rabbit and the March Hare in Lewis Carroll's Alice's Adventures in Wonderland; in the popular novels Watership Down, by Richard Adams (which has also been made into a movie) and Rabbit Hill by Robert Lawson, as well as in Beatrix Potter's Peter Rabbit stories.

Urban Legends

It was commonly believed that pregnancy tests were based on the idea that a rabbit would die if injected with a pregnant woman's urine. This is not true. However, in the 1920s it was discovered that if the urine contained the hCG, a hormone found in the bodies of pregnant women, the rabbit would display ovarian changes. The rabbit would then be killed to have its ovaries inspected, but the death of the rabbit was not the indicator of the results. Later revisions of the test allowed technicians to inspect the ovaries without killing the animal. A similar test involved injecting Xenopus frogs to make them lay eggs, but animal tests for pregnancy have been made obsolete by faster, cheaper, and simpler modern methods.

Rabbit Farming

Why Rabbit Farming?

- With available small investment and in a small place rabbit farming gives more income
- Rabbits eat ordinary feed and convert them into a protein rich high quality meat
- Apart from meat production they can also be reared for hide and fur.

Rabbit Farming is for whom?

For landless farmers, uneducated youth and women, rabbit farming gives an additional income as a part time job.

Advantages of Rabbit Farming:

- By rabbit rearing one can produce a quality protein rich meat for his own family
- Rabbits can be fed with easily available leaves, waste vegetables, grains available in the home
- Growth rate in broiler rabbits is very high. They attain 2 kgs at the age of three months
- Litter size (Number of young ones born/kindling) in rabbits is high (around 8-12)
- When compared to the other meats rabbit meat contain high protein (21%) and less fat (8%). So this meat is suitable for all age groups from adults to children.

Rabbit Breeds and Availability

Heavy weight breeds (4-6 kg):

- White giant
- Grey giant
- Flemish giant.

Medium weight breeds (3-4 kg):

- Newzealand white
- Newzealand red
- Californian.

Light weight breeds (2-3 kg):

- Soviet chinchilla
- Dutch.

Methods of Rearing Rabbits

Rabbits can be reared at the backyard with in a small shed which can be constructed with a small investment. To protect rabbits from the climatic conditions such as in hot summer, rain and to protect them from dogs, cats construction of shed is necessary.

Deep Litter System

This method is suitable for rearing small number of rabbits. To prevent rabbits to dig burrows the floor the floor should be made up of concrete. The Litter materials such as paddy husk, paddy straw or wood shavings can be filled up to 4-6 inches from the floor.

Deep litter system is not suitable for rearing more than 30 rabbits. Male rabbits should be housed separately. This type of housing is not suitable for intensive system of rearing. Rabbits reared in deep litter system are more susceptible to diseases. Management of bunnies also very difficult in deep litter system.

Cage System

Floor space requirement:

- Adult male rabbit – 4 square feet (S.ft)
- Dam – 5 S.ft
- Bunnies – 1.5 S.ft.

Adult Rabbit Cage

Adult rabbit cage should have 1.5 feet length, 1.5 feet breadth and 1.5 feet height. This cage is suitable for one adult rabbit or two growing rabbits

Growing rabbit cage:

- Length-3 feet
- Breadth-1.5 feet
- Height-1.5 feet.

The above sized cage is suitable for 4-5 rabbits up to 3 months age.

Cages for Kindling

The cage size for growing rabbits is sufficient for the rabbits going to be kindled. But the bottom and sides of the cage should be made up of weld mesh of 1.5 X 1.5 inches. This used to prevent the young bunnies to come out of the cage.

Nest Box

To provide safer and calm environment during kindling the nest box is essential. This nest boxes can be made up of galvanized iron or wood. Nest box size should be in such a way to keep inside the kindling cage.

Nest box size:

- Length – 22 inches
- Breadth-12 inches
- Height-12 inches.

Nest boxes are designed to open in the upper portion. The bottom of the nest box should be made up of weld mesh of 1.5"X 15"size. A circle size hole of 15 centimeter diameter should be made in the longitudinal portion of nest box, 10 c.m from the bottom. This hole helps for the movement of dam from the nest box to the cage. Designing of the hole 10 c.m from the bottom of the nest box prevents the young rabbits coming out of the nest box.

Cages for Rearing Rabbits in Backyard

Cages for rearing rabbits in backyard should be constructed 3'-4' from the floor surface. The bottom of the cage should be water proof.

Feed and Water Troughs

Feed and water troughs for rabbits usually made up of galvanized iron. The feed trough should be designed in the shape of "J" and they are usually fit outside the cages. To reduce the investment cost the feed and water can also be provided in cups.

Feeding Management

Rabbits relish all types of grains (sorghum, bajra and other grains) and legumes. Green fodders such as Desmanthus, Lucerne, agathi, kitchen wastes such as carrots and cabbage leaves and other vegetable wastes are also relished by rabbits.

Nutrients to be Present in the Rabbit Feed

Details of Nutrients	*For Growth*	*For Maintenance*	*For Pregnancy*	*For Lactation*
Digestible Energy (K.cal)	2500	2300	2500	2500
Protein (%)	18	16	17	19
Fibre (%)	10-13	13-14	10-13	10-13
Fat (%)	2	2	2	2

Points to be remembered in feeding management of rabbits:

- The teeth of rabbits are continuously growing. Therefore rearing rabbits with concentrate feed alone is impossible
- The time schedule for feeding should be maintained strictly. If there is late in feeding of rabbits, they become restless and which in turn reduces the body weight

- Due the high temperature in the day time the rabbits won't take feed during day time. But they active during night. Therefore the green fodder fed to rabbits at night time will helps the rabbits to consume the green fodder with out wastage. Because of this the concentrate feed should be given in the morning time
- Concentrate feed can be given in the form of pellets. If pellet feed is not available the concentrate feed mixed with water and made in to small balls and given to rabbits
- For 1 kg rabbit 40 grams of concentrate feed and 40 grams of green fodder should be given for one day
- Rabbits always fed with fresh green fodder. The green fodder should not be put on the floor of the cages but they can be inserted in the sides of the cages
- Clean fresh water should be provided to rabbits during all times of the day.

Rabbit type	*Approximate body weight*	*Amount feed/day (g)* Concentrate feed	*Green fodder*
Adult male rabbit	4-5 kg	100	250
Adult female rabbit	4-5 kg	100	300
Lactating and pregnant rabbit	4-5 kg	150	150
Young rabbits	0.6-0.7 kg	50-75	150

Sample concentrate feed mixture

Ingredients	Amount
Broken maize	30 parts
Broken and ground bajra	30 parts
Groundnut oil cake	13 parts
Wheat bran	25 parts
Mineral mixture	1.5 parts
Salt	0.5 part

Breeding Management of Rabbits

Breeding age:

- Female rabbits-5-6 months
- Male rabbits-5-6 months (Even though the male rabbits mature at the age of 5-6 months they should be used for breeding at the age of one year to get the quality younger rabbits).

Selection of rabbits for breeding:

- Rabbits can be selected at the age of 5-8 months after attaining adult body weight

- Male and female rabbits should be selected for breeding from the high litter size
- Healthy rabbits only selected for breeding. Healthy rabbits are active and the feed and water intake is usually normal. Apart from this they keep their body clean. The hair of the healthy rabbits is usually clean, soft and shiny
- While selecting male rabbits for breeding with the above characters they should have two well descended testes in their scrotum
- During selection of male rabbits they allowed to mate with a female rabbit to know the breeding capability.

Oestrous Signs or Heat Signs in Female Rabbits

There is no specific oestrous cycle period in rabbits. Whenever the female rabbits allow the male rabbit for mating then the female rabbit is in oestrous. Some times if a female rabbit is in heat, its vulva is congested. When a male rabbit is kept near to a female rabbit in heat or oestrous, the female rabbit shows depression of its back and raising of its behind portion of the body. At the same time if the female rabbit is not in heat it will go the corner of the cage and attack the male.

Breeding of Rabbits

Information about the Breeding Details

Male : female ratio	***1:10***
Age at first mating	5-6 months. Male rabbits the age of first mating usually at 1 year of age to get a good litter size
Body weight of the female rabbit during mating	2.25-2.5 kg
Gestation period	28-31 days
Weaning age	6 weeks
Time of mating after kindling	6 weeks after kindling or after weaning of the young rabbits
Age at sale	12 weeks
Body weight at the time of sale	Approximately 2 kg or greater than that

The female rabbits showing the signs of heat or oestrous taken to the male rabbit's cage. If the female rabbit is in correct time of oestrous that used to lift her tail and allow the male rabbit to mate with her. After successful mating the male rabbit fall on one side and make a specific sound. One male rabbit should not be utilised for not more than 3 or 4 days in a week. Similarly a male rabbit should not be utilised for breeding not more than 2-3 times per day. The breeding male rabbits should be given adequate rest and good nutrition. In a rabbitry there should be one male rabbit for 10 female rabbits. One or two male rabbits can also be additionally reared in the farm and they can be used if the rabbits used for breeding become sick.

Gestation period in case of broiler rabbits is 28-31 days. Pregnancy diagnosis of can be done by palpating the abdomen of the female rabbit 12-14 days after breeding. Palpation should be done in the abdominal area in between the hind legs. If there is a round mass is palpated in between the fingers then the rabbit is pregnant. Those rabbits that are not pregnant after 12-14 days of mating, they again allowed for mating with the male rabbit. If a female rabbit is not becoming pregnant after three subsequent mating, that rabbit is eliminated or culled from the farm.

Twenty five days after mating there is usually increase in the body weight of pregnant rabbits from 500-700 grams. This increased weight can be identified by lifting the rabbits. If the pregnant rabbits allowed to the male rabbit for mating they won't mate .

Care of Pregnant Female

After the pregnancy diagnosis was done, the pregnant rabbits should be fed with increased quantity of concentrate feed from 100g to 150g excess than the normal feeding. The pregnant rabbit should be transferred to the kindling cage on 25 days after mating. Five days before the expected date of mating nest box should be kept in the kindling cage. Dried coconut fibre or paddy straw is used as bedding material in the nest box. The pregnant rabbit pluck the hair in their abdomen and form a nest for the young ones one or two days before kindling. During this period the rabbit should not be disturbed and persons from outside should not allowed near the kindling cage.

Usually kindling takes place at early morning. Kindling usually completed for a period of 15 to 30 minutes. The dam her self clean her young ones in the early morning. Nest boxes should be examined in the early morning. The dead young ones should be removed from the nest box. During examination of the nest box the dam become restless. So the dam should be removed before examination of the nest box.

Care and Management of Newborn Rabbits

During birth newborn rabbits eyes are closed and they don't have hair on their body. All new born rabbits are usually lying in the bedding material formed by the dam in the nest box. Usually the dam gives milk to the young ones one time a day in the early morning. If we compulsorily make the rabbit to feed her young ones there won't be no milk secretion at all. The skin of those new born rabbits received the adequate quantity of milk from their mother is usually shiny. But the skin of those newborn rabbits who has not received the adequate quantity of milk from their mother is dry and wrinkled in appearance and their body temperature is low and they look lazy.

Step Mother Feeding

Generally a female rabbit has 8-12 teats in her udder. When the litter size is more than the number of teats the newborn rabbits won't receive adequate quantity of milk and results in death of the young ones. In an addition the other conditions such as death of dam, lack of maternal care, falling of young ones from the cage results in difficulty in the identification of the dams, the step mother is used for nursing the young ones.

Points to be considered while changing the litter to the step mother:

- The age difference between the litters to be changed and the litter of step mother should not be more than 48 hours
- We should not change more than 3 young ones/stepmother.

Weaning of Bunnies or Young Rabbits

Young rabbits allowed staying for first three weeks of age in the nest box. Later nest box removed from the kindling cage. Weaning of young rabbits or bunnies can be done at the age of 4-6 weeks. While weaning, first dam should be removed from the kindling cage and the bunnies allowed staying for 1-2 weeks in the same cage. Later the sex of the rabbits should be identified and the different sexes are raised in different cages. We should not suddenly change the diet or feed of the weaned rabbits.

Reduction of Mortality Rate in the Bunnies

Up to first 15 days of age the young rabbits are under the dam. During this period the breast milk of the doe or dam is the only food for the young ones. Death of the young rabbits in this period is mainly due to doe or dam. After 15 days of age the young rabbits able to take water and feed supplied to them. In this time they are more susceptible to diseases. Therefore it is advisable to provide cooled boiled drinking water to the dam and the young ones. Hydrogen peroxide at the rate of 1ml per liter of water is added 20 minutes before providing to the rabbits.

Signs of healthy rabbit:

- Healthy and shiny hair coat
- very active
- Good and quick feed intake after feeding
- The eyes are usually bright and without any discharge
- Gradual body weight gain.

Signs in diseased rabbits:

- Dull and depressed
- Reduction in the body weight and emaciation
- Heavy loss of hair

- No active movement of rabbits. But usually they stay in one particular place in the cage
- Reduction of feed intake
- Watery or mucous discharge from the eye, nose, anus and mouth
- Increased body temperature and respiratory rate.

Diseases of Rabbits

Pasteurellosis

Poor ventilation, poor hygiene, nutritional deficiency are the risk factors responsible for occurrence of this disease in rabbits. This disease spreads from dam to young rabbits.

Clinical signs: Because of continuous sneezing and coughing the rabbits rub their nose with the forelimbs. Sound produced during breathing resembles like the rattle sound. Apart from this there is also fever, diarrhea. The microorganism responsible for this disease also produces pustules under the skin and wry neck in rabbits.

Treatment : Treatment is not effective against pasteurellosis. Even though the rabbits affected with pasteurellosis recover after treatment but the affected rabbits transmit the infectious bacteria to other healthy rabbits. Therefore culling the affected rabbits from the farm is the only control measure of pasteurellosis in the rabbitry.

Enteritis

There are various microorganisms responsible for producing enteritis in rabbits. Sudden change in the feed, excess amount of carbohydrate in the feed, immuno depression, unhygienic feed and water are the factors predispose the rabbits for enteritis caused by various microorganisms. The clinical signs of enteritis in rabbits are diarrhea, enlargement of abdomen, dull hair coat and dehydration. Due to the loss of water due to diarrhoea which leads to dehydration, rabbits become dull.

Wry Neck Disease

Rabbits affected with the pasteurellosis are affected with wry neck disease. This disease affects the middle ear and the brain of rabbits. Because of the affection of membrane of the middle ear and pus is discharged from the ear, the rabbit turn its head to one side. Effective treatment of pasteurellosis controls the wry neck disease in rabbits.

Mastitis

Nursing dams are affected with mastitis. Affected udder is hot, reddish and painful to touch. Giving appropriate antibiotics to rabbits will control the disease.

Diseases Due to Fungal Infection

Skin infection is produced in rabbits by Dermatopysis fungus. There is loss of hair around the ear and nose of rabbits. Because of irritation rabbits continously rub the affected areas results in formation of wounds in these areas. Later on secondary bacterial infection in these sites results in pus formation.

Treatment : Griseofulvin or benzyl benzoate cream can be applied on the affected areas. Griseofulvin mixed at the rate of 0.75g/k.g of feed and given for a period of 2 weeks control this disease.

Farm hygienic measures to be followed to control disease in rabbitry:

- Rabbit farm should be located in elevated place with good ventilation
- The cages should be kept very clean
- There should be trees around the rabbit shed
- White washing should done twice a year
- Lime solution should be applied below the cages twice a week
- In summer seasons water may be sprinkled on the rabbits to avoid death due to heat stroke
- Before providing drinking water it should be boiled and then cooled especially to the dam and young ones
- To prevent diseases caused by bacterial microorganisms, Tetracycline mixed with drinking water at the rate of 0.5 g/litter and given for a period of 3 days per month.

5

Piggery Management

Domestic Pig

The domestic pig, known in some areas as the hog, is a domesticated livestock animal, farmed for meat (pork, bacon, ham, sausages, etc.)

Most domestic pigs have rather sparse hair covering on their skin, although woolly coated breeds are known (Mangalitsa pig), and some were popular in the past.

The domestic pig is most often treated as a subspecies of its wild ancestor, the wild boar, and in this case it is given the scientific name Sus scrofa domesticus. Some taxonomists treat the domestic pig as a separate species, when it is called Sus domesticus, and wild boar is S. scrofa. Wild boar were in human association as early as 13,000–12,700 BP. Escaped domestic pigs have become feral in many parts of the world (for example, New Zealand) and have caused substantial environmental damage.

Origins

Archeological evidence suggests that pigs were domesticated from wild boar as early as 13,000–12,700 BP in the Near East in the Tigris Basin being managed in the wild in a way similar to the way they are managed by some modern New Guineans. Remains of pigs have been dated to earlier than 11,400 BP in Cyprus that must have been introduced from the mainland which suggests domestication in the adjacent mainland by then. There was also a separate domestication in China..

DNA evidence from sub-fossil remains of teeth and jawbones of Neolithic pigs in Europe shows that the first domestic pigs there had been brought from the Near East. This stimulated the domestication of local European wild boar resulting in a third domestication event with the Near Eastern genes ceasing in European pig stock. Modern domesticated pigs have involved complex exchanges with European domesticated lines being exported in turn to the ancient Near East. Historical records indicate that Asian pigs were introduced into Europe during the 18th and early 19th centuries.

The adaptable nature and omnivorous diet of the wild boar allowed early humans to domesticate it readily. Pigs were mostly used for food, but early civilizations also used the pigs' hides for shields, bones for tools and weapons, and bristles for brushes. Pigs were brought to southeastern North America from Europe by de Soto and other early Spanish explorers. Escaped pigs became feral and caused a great deal of disruption to Native Americans cultures who had no domesticated livestock.

Food

The domestic pig is used for its meat, called pork. Other products made from pigs include sausage, bacon, gammon, ham, head cheese, and pork scratchings. The head of a pig can be used to make a preserved jelly called head cheese (brawn). Liver, chitterlings, blood (blood pudding or black pudding) and other offal from pigs are also widely used for food. In some religions, such as Judaism and Islam, there are religious restrictions on the consumption of pork.

Farming

In developing nations, and often in developed nations, the domestic pig is raised outdoors in yards or fields. In some cases pigs are allowed to forage in woods, where they may be watched by swineherds, the equivalent of shepherds for pigs. In industrialized nations, domestic pig farming has shifted away from the traditional pig farm to large-scale intensive pig farms where meat can be mass-produced. This has resulted in lower production costs, but more significant animal welfare concerns.

Individual farm management practices focus on, among other things, housing facilities, feeding and ventilation systems, and temperature and environmental controls. Just as producers have to determine the type of facilities and equipment for their farm, they must find the practices that best fit their farm's specific situation. Some procedures and treatments can be short-term stressors, so producers must weigh the long-term welfare, health and management benefits to the animals.

Using the knowledge obtained from scientific study, some producers have adopted and adapted techniques and husbandry skills to protect the welfare of their animals. They feel a personal and moral responsibility to take care of their animals and ensure that the animals are safe, but they also must earn a profit from their business. They consider anything short of providing the best, humane care possible as being self-defeating.

Uses in Truffle Hunting

The domestic pig is also widely used in France and other countries to search for truffles. This is because the truffle smells similar to the pheromones of the animal.

As Pets

Pigs are known to be intelligent animals and can be trained similarly to dogs, though they may excel in different tasks. Asian pot-bellied pigs, a small type of domestic pig, have made popular house pets in the United States beginning in the latter half of the 20th century.

Regular domestic farmyard pigs have also been known to be kept indoors, but due to their large size and destructive tendencies, they typically need to be moved into an outdoor pen as they grow older. Most pigs have a fear of being picked up, but will usually calm down once placed back on the floor. Pigs are rarely used as working animals. An exception is the use of truffle pigs – ordinary pigs trained to find truffles.

Miniature pigs, also called micro or teacup pigs, which are specifically bred to be small (from 29-65 lbs) gained in popularity in Fall of 2009 after several mainstream press articles claimed they were a popular pet to celebrities such as Rupert Grint of Harry Potter fame.

Breeds of Pigs

Pigs are exhibited at agricultural shows, judged either as stud stock compared to the standard features of each breed, or in commercial classes where the animals are judged primarily on their suitability for slaughter to provide premium meat.

According to the American Livestock Breeds Conservancy, seven breeds of swine in the U.S. are critically rare (having a global population of fewer than 2000). Outside the U.S., the Auckland Island Pig (New Zealand) and woolly-coated grazing pig (Danube area) are known to be critically rare.

Pig Farming Terminology

Types of animal:

- Pig hog or swine, the species as a whole, or any member of it. The singular of "swine" is the same as the plural.
- Shoat, piglet or (where the species is called "hog") pig, unweaned young pig, or any immature pig.
- Sucker, a pig between birth and weaning.
- Runt, an unusually small and weak piglet, often one in a litter.
- Boar or hog, male pig of breeding age.
- Barrow, male pig castrated before puberty.
- Stag, male pig castrated later in life, (that is, an older boar after castration).
- Gilt, young female not yet mated, or not yet farrowed, or after only one litter (depending on local usage).
- Sow, breeding female, or female after first or second litter.

Pigs for Slaughter:

- suckling pig, a piglet slaughtered for meat.
- Feeder pig, a weaned gilt or barrow weighing between 18 kg (40 lb) and 37 kg (82 lb) at 6 to 8 weeks of age that is sold to be finished for slaughter.
- Porker, market pig between 30 kg (66 lb) and about 54 kg (120 lb) dressed weight.
- Baconer, a market pig between 65 kg (140 lb) and 80 kg (180 lb) dressed weight. The maximum weight can vary between processors.
- *Grower, a pig between weaning and sale or transfer to the breeding herd, sold for slaughter or killed for rations.*
- Finisher, a grower pig over 70 kg (150 lb) liveweight.
- Butcher hog, a pig of approximately 100 kg (220 lb), ready for the market.

·Backfatter, cull breeding pig sold for meat; usually refers specifically to a cull sow, but is sometimes used in reference to boars.

Groups:

- Herd, a group of pigs, or all the pigs on a farm or in a region.
- Sounder, a small group of pigs (or wild boar) foraging in woodland.

Pig Parts:

- Trotters, the feet of pigs (they have four hoofed toes, walking mainly on the larger central two).

Biology:

- In pig, pregnant.
- Farrowing, giving birth.
- Hogging, a sow when on heat (during oestrus).

Housing:

- Sty, a small pig-house, usually with an outdoor run or a pig confine meant.
- Pig-shed, a larger pig-house.
- Ark, a low field-shelter for pigs (or other animals such as rabbits or chickens).
- Curtain-barn, a long, open building with curtains on the long sides of the barn. This increases ventilation on hot, humid summer days.

Pig Farming

Advantages of pig rearing:

- Pigs convert inedible feeds, forages, certain grain by-products obtained from mills, meat by products, damaged feeds and garbage into valuable nutritious meat. Most of these feeds are either not edible or not very palatable to human beings

- Pig grows fast and is a prolific breeder, farrowing 10 to 12 piglets at a time. It is capable of producing two litters per year under optimal management conditions
- The carcass return is quite high ie. 60-80 percent of live body weight
- With a small investment on building and equipment, proper feeding and sound disease control programme the farmer can profitably utilize his time and labour in this subsidiary occupation
- The faeces of pigs is used as a manure to maintain soil fertility.

Pig farming-for whom?

- Small and landless farmers
- Part time earning for educated youth having agriculture as occupation
- uneducated youth
- Farm women.

Breeds

The indigenous pig has been the basis used for pig production for a long period of time. It is small in size. Improved breeds are now being used for grading up the form the basis for pig production in the rural areas.

Large white Yorkshire:

- Most extensively used exotic breed in India
- Body colour is solid white with occasional black – pigmented spots
- Erect ears, snout of medium lengths and dished face
- Excellent breed for the purpose of cross breeding
- Prolific breeds
- Mature boar 300-400 k.g
- Mature sow 230-320 kg.

Landrace:

- White with black skin spot
- Long body, large drooping ears and long snout
- Prolific breeding and efficient in utilizing feed
- Carcass quality equal to Yorkshire
- Excellent breed for crossbreeding
- Mature boars weigh 270-360 kg
- Mature sow weighs 200-320 kg.

Middle white Yorkshire:

- Used in some areas in India

- Grows rapidly gives good dressing percentage
- Not so prolific as large white Yorkshire
- Males 250-340 kg
- Females 180-270 kg.

Breeds for North East India

Pig farming is very important component in North East India. Out of total pig population in India, 28% are grown in this region.

Selection of Breeding Stock

Important characteristics which need to be considered in developing a good sow herd are

- Size of litters
- Strength and vigour of litters
- Milking ability
- Temperament.

Gain and feed efficiency, fertility, selection of individual animals from a herd is more important than the selection of a particular breed. Each producer at the time of setting up his herd should purchase his animals from a reliable disease free herd and should obtain as much information of the animals as possible. Once the herd is established the selection of the gilts and boars for replacement in the breeding herd should be based on the types and performance.

Yorkshire gilt (female):

- Selection of gilts for the breeding herd should be made at market weight ie. When the animals weigh about 90 kg
- Select gilts from sows which have consistently farrowed and weaned large litters
- They have reached market in minimum time and have desirable market type
- It would be desirable to choose gilts whose litter mates and other fullsibs have given good performance in daily weight gain and feed conversion efficiency.

Yorkshire boar (male):

- Selection of boar is extremely important, particularly for a small breeding farm or unit
- The boar should be purchased from a breeder or a farm maintaining adequate information on its performance
- The boar should be selected from a dam which has consistently farrowed and weaned high litters

- A good boar will weigh 90 kg in about 5-6 months will be of good type and will be strong on feet and legs
- The feed conversion from weaning to 90 kg weight would be the most desirable.

Points to be considered while replacing boars and gilts:

- The mother of the pig to be selected should have had large litters of 8 piglets or more. The weaning weight (at 56 days) of a litter in case of a gilt selection should have been 120 kg and in the case of sow it should not less than 150 kg
- The gilt or the boar should have reached a body weight of about 90 kg in about 6 months
- The pig should have adequate length and depth of the body, thick well muscled hams should be firm and trim
- The pig should have sound feet and legs
- The back fat prove in live animals is not yet being practiced in this country. For gilt back fat thickness of 4 cm or less, boars 3.2 or less
- Gilts should have a minimum of 12 evenly spaced, functional teats. An animal with blind teats should be avoided as there will be little or no milk from these teats and the defect is heritable
- Negative blood test for both brucellosis and leptospirosis should be made during selection and the pigs should be vaccinated against swine fever
- Pigs should be free from other diseases and physical defects.

Age of Breeding Stock

Well developed gilts may as a general rule bred to farrow when 12-14 months old. This depends more on development than on age. Gilts should weigh at least 100 kg before breeding. Ovulation rate increases during successive oestrous periods (up to fifth) following puberty. Thus it is advantageous to delay the breeding of gilts until the second or third oestrous. Litter size increases on an average in succeeding pregnancies up to 5th or 6th litter. It is therefore advantageous to cull the sow from a breeding herd or a commercial herd after her fifth or sixth litter as the litter size goes down thereafter.

Detection of Heat

The average length of oestrous cycle in pigs is 21 days. The oestrous symptoms last for five to seven days beginning with vulvar swelling and vaginal discharge. In true oestrous there is frequent urination, reduced appetite, mounting and standing for service detected by the erection of ears and immobility when normal pressure is applied to the back. The application of pressure on the back is used to determine the correct breeding time. Animals with a predisposition for weak oestrous should be brought near the boar to exhibit heat symptoms a little more clearly.

Best time for breeding is during the latter half of the first day or early on the second day of oestrous. In many cases the gilts and sows continue to exhibit the standing heat on the next day. In these cases the animals should be rebred and the interval in the case of rebreeding should be 12-14 hours. This procedure will ensure a high conception rate in the herd.

Sows come into oestrous one to four days after farrowing but they should not be bred at this time. Sows may also come into heat two to ten days after weaning and may be bred at this time. But better results can be obtained by breeding them in the second post lactational oestrous. The animals which have been bred should be observed for the appearance of subsequent oestrous. If sows not conceived even after successful mating with a boar in two continous oestrous cycles it is desirable to cull them from the herd.

Flushing

It is the method of feeding sows and gilts before breeding. A good grower ration fed to sows and gilts seven to ten days before breeding is helps in increased ovulation rates in them. After breeding sows and gilts should be fed a limited but well balanced ration until the last six weeks of pregnancy and then full feeding should be resumed.

Care and Management of Pregnant Animals

The gestation period of sow varies from 109-120 days with an average of 114 days. Pregnant animals should be housed in groups in separate enclosures and should not be mixed with new animals to avoid fighting which at times may result in abortion. It would also be advisable to house pregnant gilts and sows in separate groups during gestation. About 3 m2 of dry housing should be available for each sow. The pregnant animals should be allowed to move about every day in the morning on a free range or a pasture if available. A pasture area is presumed to be clean if a cultivated crop was raised.

Management at Farrowing

Farrowing time is the critical time in pig production. Death rate is high during farrowing and the first week after farrowing. Sows may be farrowed in pens equipped with guard rails and a creep space in farrowing crates or in farrowing stalls. A pen equipped with guard rails and a creep space is adequate. The pen should be maintained at 24°C to 28°C until the piglets are three or four days old and at 18°C to 22°C until the piglets are approximately six weeks old. The heat lamps should be hung 45 cm from the floor and suitably protected. The farrowing pens should be thoroughly cleaned before the sow is brought in. This will prevent a large number of diseases of piglets. The sow should brought to the farrowing pen atleast one week prior to farrowing so that it becomes familiar with the surroundings. She should be washed thoroughly before being brought to the farrowing pen. The feed ration should be made bulky by substituting one-third of the regular ration with wheat bran. The amount of ration fed should also be reduced by one third till the sow farrows. The sow should be watched closely for determining the approximate time of farrowing and feed should not be given 12 hours before farrowing.

Care During Farrowing

An attendant should be on hand when the sow farrows. Otherwise many piglets will die. It takes generally 2 to 4 hours for complete farrowing to take place. The piglets should be removed as they are farrowed and kept warm in the creep space until farrowing is complete. Each piglet should be cleaned of all mucus to ensure that the breathing passages are clear. The navel cord should be tied 2-5 cm away from the navel, cut with a disinfected pair of scissors and the stumps painted with iodine. Piglets should be allowed to be nursed after birth. In about 2 days they settle down to their individual teats. They nurse 8-10 times in 24 hours in the initial period. Trampling by the sow should be prevented during the first two weeks.

Removal of Needle Teeth

Piglets are born with four pairs of sharp teeth, with two pairs on each jaw. They are of no practical value to the piglets and they may irritate the sow's udder during nursing or cause injury to other piglets. Clipping of these teeth shortly after birth will prevent the injury of the udder caused by the needle teeth.

Anaemia in Piglets

Anaemia is a common nutritional disease in piglets. This condition can be prevented and cured by supplying iron either orally or by injection. Oral administraion consists of spraying or swabbing the sow's udder with a saturated solution of ferrous sulphate (0.5 kg of ferrous sulphate in 10 litres of hot water). This solution must be applied daily from birth until the piglets start eating creep feed. Intra muscular infection of iron –dextran compounds is the more effective method of preventing anaemia.

Raising Orphan Piglets

The death of a sow after farrowing, mastitis, lactation failure of litters larger than the sow is able to raise, result in orphan pigs. If another sow has farrowed within a short time previously, the orphan piglets may be transferred to her. This transfer must be made within a few days after farrowing because those section of sow's udder are not used soon cease producing milk. To ensure acceptance of new pigs the sow should be separated from her own litter for short time and then the new piglets are brought to her and a disinfectant or other material sprinkled on all the piglets to mask the odours.

Orphan piglets can also be raised with milk replacer. Milk replacer consists of one egg yolk thoroughly mixed with one litre of cow milk. This mixture supplies a well balanced diet except for iron. To compensate for the lack of iron one eighth teaspoonfull of ferrous sulphate may be added to one litre of milk. An injection with iron compound may also be used.

Castration

The male piglets not selected for breeding may be castrated when they are three to four weeks old.

Management During Lactation

Sows and gilts should be fed with good judgment during lactation. When fully fed a sow needs about 0.5 kg of feed per piglet she is nursing in addition to her own requirements.

Separation of Piglets from Mother (Weaning)

Normal weaning age of piglets is at 8 weeks age. The sow should be separated from the piglets for a few hours each day to prevent stress of weaning and feed is reduced gradually. The piglets should be dewormed after 2 weeks of weaning. The piglets should be gradually shifted from 18 percent protein creep feed to 16 percent grower ration over a period of two weeks. Group of 20 piglets of more or less the same age should be housed in each pen.

Prevention and control of swine diseases:

- All pigs should be vaccinated against swine fever at the age of 2-4 weeks. Breeding pigs should be tested for brucellosis and leptospirosis. As a routine measure all young pigs at the time of weaning should be inoculated against swine fever.
- Animals purchased for the farm should be purchased from disease free herds. Newly purchased animals should be isolated from the other animals in the farm for a period of three to four weeks. No visitor allowed visiting the farm. Those stys or pig houses cleared of the animals are kept empty for three to four weeks for destruction of microorganisms causing the disease.

6

Aquaculture

Aquaculture is the farming of freshwater and saltwater organisms such as finish, mollusks, crustaceans and aquatic plants. Also known as aquafarming, aquaculture involves cultivating aquatic populations under controlled conditions, and can be contrasted with commercial fishing, which is the harvesting of wild fish. Commercial aquaculture supplies one half of the fish and shellfish that is directly consumed by humans.

Mariculture refers to aquaculture practiced in marine environments. Particular kinds of aquaculture include algaculture (the production of kelp/seaweed and other algae), fish farming, shrimp farming, oyster farming, the growing of cultured pearls and the growing and selling of ornamental fish. Particular methods include aquaponics, which integrates fish farming and plant farming.

History

The indigenous Gunditjmara people in Victoria, Australia may have raised eels as early as 6000 BC. There is evidence that they developed about 100 square kilometres (39 sq mi) of volcanic floodplains in the vicinity of Lake Condah into a complex of channels and dams, that they used woven traps to capture eels, and that capturing and smoking eels supported them year round.

Aquaculture was operating in China circa 2500 BC. When the waters subsided after river floods, some fishes, mainly carp, were trapped in lakes. Early aquaculturists fed their brood using nymphs and silkworm feces, and ate them. A fortunate genetic mutation of carp led to the emergence of goldfish during the Tang Dynasty.

Japanese cultivated seaweed by providing bamboo poles and, later, nets and oyster shells to serve as anchoring surfaces for spores.

In central Europe, early Christian monasteries adopted Roman aquacultural practices. Aquaculture spread in Europe during the Middle Ages, since away from the seacoasts and the big rivers, fish were scarce/expensive. Improvements in transportation during the 19th century made fish easily available and inexpensive, even in inland areas, making aquaculture less popular.

Hawaiians constructed oceanic fish ponds. A remarkable example is a fish pond dating from at least 1,000 years ago, at Alekoko. Legend says that it was constructed by the mythical Menehune dwarf people.

In 1859 Stephen Ainsworth of West Bloomfield, New York, began experiments with brook trout. By 1864 Seth Green had established a commercial fish hatching operation at Caledonia Springs, near Rochester, New York. By 1866, with the involvement of Dr. W. W. Fletcher of Concord, Massachusetts, artificial fish hatcheries were under way in both Canada and the United States. When the Dildo Island fish hatchery opened in Newfoundland Canada in 1889, it was the largest and most advanced in the world.

Californians harvested wild kelp and attempted to manage supply circa 1900, later labeling it a wartime resource.

In the 21st century

About 430 (97%) of the species cultured as of 2007 were domesticated during the 20th century, of which an estimated 106 came in the decade to 2007. Given the long-term importance of agriculture, it is interesting to note that to date only 0.08% of known land plant species and 0.0002% of known land animal species have been domesticated, compared with 0.17% of known marine plant species and 0.13% of known marine animal species. Domestication typically involves about a decade of scientific research.

Domesticating aquatic species involve fewer risks to humans than land animals, which took a large toll in human lives. Most major human diseases originated in domesticated animals. through diseases such as smallpox and diptheria, that like most infectious diseases, move to humans from animals. No human pathogens of comparable virulence have yet emerged from marine species.

Harvest stagnation in wild fisheries and overexploitation of popular marine species, combined with a growing demand for high quality protein encourages aquaculturists to domesticate other marine species.

Production Volume

In 2004, the total world production of fisheries was 140,500,000 tonnes (138,300,000 LT; 154,900,000 ST) of which aquaculture contributed 45,500,000 tonnes (44,800,000 LT; 5.02E+7 ST) or about 32%. The growth rate of worldwide aquaculture has been sustained and rapid, averaging about 8 percent per annum for over thirty years, while the take from wild fisheries has been essentially flat for the last decade.

Top Ten Species Groups in 2004

Species group	***Million tonnes***
Carps and other cyprinids	18.30
Oysters	4.60
Clams, cockles, ark shells	4.12
Miscellaneous freshwater fishes	3.74
Shrimps, prawns	2.48
Salmons, trouts, smelts	1.98
Mussels	1.86
Tilapias and other cichlids	1.82
Scallops, pectens	1.17
Miscellaneous marine molluscs	1.07

Aquaculture is an especially important economic activity in China. Between 1980 and 1997, the Chinese Bureau of Fisheries reports, aquaculture harvests grew at an annual rate of 16.7 percent, jumping from 1,900,000 tonnes (1,870,000 LT; 2,090,000 ST) to nearly 23,000,000 tonnes (23,000,000 LT; 25,000,000 ST). In 2005, China accounted for 70% of world production. It is currently one of the fastest growing areas of food production in the U.S.

Top Ten Aquaculture Producers in 2004

Country	***Million tonnes***
China	30.61
India	2.47
Viet Nam	1.20
Thailand	1.17
Indonesia	1.05
Bangladesh	0.91
Japan	0.78
Chile	0.67
Norway	0.64
United States	0.61
Other countries	5.35
Total	45.47

Approximately 90% of all U.S. shrimp consumption is farmed and imported. In recent years salmon aquaculture has become a major export in southern Chile, especially in Puerto Montt, Chile's fastest-growing city.

Over Reporting

China overwhelmingly dominates the world in reported aquaculture output. They report a total output which is double that of the rest of the world put together. However, there are issues with the accuracy of China's returns.

In 2001, the fisheries scientists Reg Watson and Daniel Pauly expressed concerns in a letter to Nature, that China was over reporting its catch from wild fisheries in the 1990s. They said that made it appear that the global catch since 1988 was increasing annually by 300,000 tonnes, whereas it was really shrinking annually by 350,000 tonnes. Watson and Pauly suggested this may be related to China policies where state entities that monitor the economy are also tasked with increasing output. Also, until recently, the promotion of Chinese officials was based on production increases from their own areas.

China disputes this claim. The official Xinhua News Agency quoted Yang Jian, director general of the Agriculture Ministry's Bureau of Fisheries, as saying that China's figures were "basically correct". However, the FAO accepts there are issues with the reliability of China's statistical returns, and currently treats data from China, including the aquaculture data, apart from the rest of the world.

Methods

Mariculture: Mariculture is the term used for the cultivation of marine organisms in seawater, usually in sheltered coastal waters. In particular, the farming of marine fish is an example of mariculture, and so also is the farming of marine crustaceans (such as shrimps), molluscs (such as oysters) and seaweed.

Integrated: Integrated Multi-Trophic Aquaculture (IMTA) is a practice in which the by-products (wastes) from one species are recycled to become inputs (fertilizers, food) for another. Fed aquaculture (e.g. fish, shrimp) is combined with inorganic extractive (e.g. seaweed) and organic extractive (e.g. shellfish) aquaculture to create balanced systems for environmental sustainability (biomitigation), economic stability (product diversification and risk reduction) and social acceptability (better management practices).

"Multi-Trophic" refers to the incorporation of species from different trophic or nutritional levels in the same system. This is one potential distinction from the age-old practice of aquatic polyculture, which could simply be the co-culture of different fish species from the same trophic level. In this case, these organisms may all share the same biological and chemical processes, with few synergistic benefits, which could potentially lead to significant shifts in the ecosystem. Some traditional polyculture systems may, in fact, incorporate a greater diversity of species, occupying several niches, as extensive cultures (low intensity, low management) within the same pond. The "Integrated" in IMTA refers to the more intensive cultivation of the different species in proximity of each other, connected by nutrient and energy transfer through water. Ideally, the biological and chemical processes in an IMTA system should balance. This is achieved through the appropriate selection and proportions of different species providing different ecosystem functions. The co-cultured species are typically more than just biofilters; they are harvestable crops of commercial value. A working IMTA system can result in greater total production based on mutual benefits to the co-cultured species and improved ecosystem health, even if the production of individual species is lower than in a monoculture over a short term period.

Sometimes the term "Integrated Aquaculture" is used to describe the integration of monocultures through water transfer. For all intents and purposes however, the terms "IMTA" and "integrated aquaculture" differ only in their degree of descriptiveness. Aquaponics, fractionated aquaculture, IAAS (integrated agriculture-aquaculture systems), IPUAS (integrated peri-urban-aquaculture systems), and IFAS (integrated fisheries-aquaculture systems) are other variations of the IMTA concept.

Species Groups

Finfish

The farming of finfish is the most common form of aquaculture. It involves raising fish commercially in tanks, ponds, or ocean enclosures, usually for food. A facility that releases juvenile fish into the wild for recreational fishing or to supplement a species' natural numbers is generally referred to as a fish hatchery. Fish species raised by fish farms include salmon, bigeye tuna, carp, tilapia, catfish and cod.

In the Mediterranean, young bluefin tuna are netted at sea and towed slowly towards the shore. They are then interned in offshore pens where they are further grown for the market. In 2009, researchers in Australia managed for the first time to coax tuna (Southern bluefin) to breed in landlocked tanks.

Shellfish

Abalone farming began in the late 1950s and early 1960s in Japan and China. Since the mid-1990s, this industry has become increasingly successful. Over-fishing and poaching have reduced wild populations to the extent that farmed abalone now supplies most abalone meat.

Crustaceans

Commercial shrimp farming began in the 1970s, and production grew steeply thereafter. Global production reached more than 1,600,000 tonnes (1,570,000 LT; 1,760,000 ST) in 2003, representing a value of nearly 9,000 million U.S. dollars. About 75% of farmed shrimp is produced in Asia, in particular in China and Thailand. The other 25% is produced mainly in Latin America, where Brazil is the largest producer. Thailand is the largest exporter.

Shrimp farming has changed from its traditional, small-scale form in Southeast Asia into a global industry. Technological advances have led to ever higher densities per unit area, and broodstock is shipped worldwide. Virtually all farmed shrimp are penaeids (i.e., shrimp of the family Penaeidae), and just two species of shrimp—the Penaeus vannamei (Pacific white shrimp) and the Penaeus monodon (giant tiger prawn) account for roughly 80% of all farmed shrimp. These industrial monocultures are very susceptible to disease, which has decimated shrimp populations across entire regions. Increasing ecological problems, repeated disease outbreaks, and pressure and criticism from both NGOs and consumer countries led to changes in the industry in the late 1990s and generally stronger regulations. In 1999, governments, industry representatives, and environmental organizations initiated a program aimed at developing and promoting more sustainable farming practices.

Freshwater prawn farming shares many characteristics with, including many problems with marine shrimp farming. Unique problems are introduced by the developmental life cycle of the main species (the giant river prawn, Macrobrachium rosenbergii).

The global annual production of freshwater prawns (excluding crayfish and crabs) in 2003 was about 280,000 tonnes (280,000 LT; 310,000 ST) of which China produced 180,000 tonnes (180,000 LT; 200,000 ST) followed by India and Thailand with 35,000 tonnes (34,000 LT; 39,000 ST) each. Additionally, China produced about 370,000 tonnes (360,000 LT; 410,000 ST) of Chinese river crab (Eriocheir sinensis).

Echinoderms

Commercially harvested echinoderms include sea cucumbers and sea urchins. In China, sea cucumbers are farmed in artificial ponds as large as 1,000 acres (400 ha).

Algae

Microalgae, also referred to as phytoplankton, microphytes, or planktonic algae constitute the majority of cultivated algae.

Macroalgae, commonly known as seaweed, also have many commercial and industrial uses, but due to their size and specific requirements, they are not easily cultivated on a large scale and are most often taken in the wild.

Issues

Aquaculture can be more environmentally damaging than exploiting wild fisheries. Concerns include waste handling, side-effects of antibiotics, competition between farmed and wild animals, and using other fish to feed more marketable carnivorous fish. However, research and commercial feed improvements during the 1990s & 2000s have lessened many of these. Fish waste is organic and composed of nutrients necessary in all components of aquatic food webs. In-ocean aquaculture often produces much higher than normal fish waste concentrations. The waste collects on the ocean bottom, damaging or eliminating bottom-dwelling life. Waste can also decrease dissolved oxygen levels in the water column, putting further pressure on wild animals.

Cultivators often supply their animals with antibiotics to prevent disease. As with livestock, this can accelerate the evolution of bacterial resistance.

Fish Oils

The nutritional value of farm-raised tilapia may be compromised due to the amount of corn included in the feed. Corn contains short chain omega-6 fatty acids that contribute to the buildup of these materials in the fish. "Ratios of long-chain omega-6 to long-chain omega-3, AA to EPA respectively, in tilapia averaged about 11:1, compared to much less than 1:1 (indicating more EPA than AA) in both salmon and trout." The US produced 1,500,000 tonnes (1,480,000 LT; 1,650,000 ST) of tilapia in 2005, with 2,500,000 tonnes (2,460,000 LT; 2,760,000 ST) projected by 2010. Widespread publicity encouraging fish consumption has led to increases in tilapia consumption by those with lower incomes who are trying to eat a balanced diet. The lower amounts of omega-3 and the higher ratios of omega-6 compounds in farmed tilapia raise questions of the health benefits of consuming this fish.

Adequate diets for salmon and other carnivorous fish can be formulated from protein sources such as soy, although soy-based diets may also change in the balance between omega-6 and omega-3 fatty acids.

Impacts on Wild Fish

Salmon farming currently involves a high demand for wild forage fish for feed. As carnivores, salmon need a lot of protein, and farmed salmon eat more fish than they produce. Each pound of farmed salmon require up to 6 pounds of wild fish. Seventy five percent of the worlds monitored fisheries are already near to or have exceeded their maximum sustainable yield." The industrial scale extraction of wild forage fish for salmon farming also impacts the survivability of the wild predator fish who rely on them for food.

Fish can escape from coastal pens, where they can interbreed with their wild counterparts, diluting wild genetic stocks. Escaped fish can become invasive, outcompeting native species.

Coastal Ecosystems

Aquaculture is becoming a significant threat to coastal ecosystems. About 20 percent of mangrove forests have been destroyed since 1980, partly due to shrimp farming. Large scale conversions of mangroves into brackish shrimp ponds have been characterized as the marine equivalent of "slash-and-burn" farming. An extended cost–benefit analysis of the total economic value of shrimp aquaculture built on mangrove ecosystems found that the external costs were much higher than the external benefits. Over four decades, 269,000 hectares (660,000 acres) of Indonesian mangroves have been converted to shrimp farms. Most of these farms are abandoned within a decade because of the toxin build-up and nutrient loss.

Salmon farms are typically sited in pristine coastal ecosystems which they then pollute. A farm with 200,000 salmon discharges more fecal waste than a city of 60,000 people. This waste is discharged directly into the surrounding aquatic environment, untreated, often containing antibiotics and pesticides." There is also an accumulation of heavy metals on the benthos (seafloor) near the salmon farms, particularly copper and zinc.

Genetic Modification

Salmon have been genetically modified for faster growth, although they are not approved for commercial use, in the face of opposition. One study, in a laboratory setting, found that modified salmon mixed with their wild relatives were aggressive in competing, but ultimately failed.

Prospects

Global wild fisheries are in decline, with valuable habitat such as estuaries in critical condition. The aquaculture or farming of piscivorous fish, like salmon, does not help the problem because they need to eat products from other fish, such as fish meal and fish oil. Studies have shown that salmon farming has major negative impacts on wild salmon, as well as the forage fish that need to be caught to feed them. Fish that are higher on the food chain are less efficient sources of food energy.

Apart from fish and shrimp, some aquaculture undertakings, such as seaweed and filter-feeding bivalve mollusks like oysters, clams, mussels and scallops, are relatively benign and even environmentally restorative. Filter-feeders filter pollutants as well as nutrients from the water, improving water quality. Seaweeds extract nutrients such as inorganic nitrogen and phosphorus directly from the water, and filter-feeding mollusks can extract nutrients as they feed on particulates, such as phytoplankton and detritus. Some profitable aquaculture cooperatives promote sustainable practices. New methods lessen the risk of biological and chemical pollution through minimizing fish stress, fallowing netpens, and applying Integrated Pest Management. Vaccines are being used more and more to reduce antibiotic use for disease control. Onshore recirculating aquaculture systems, facilities using polyculture techniques, and properly-sited facilities (e.g. offshore areas with strong currents) are examples of ways to manage negative environmental effects.

Fishing in India

Fish production has increased more than fivefold since India's independence and is a major industry in the coastal states.

Growth

It rose from only 800,000 tons in FY 1950 to 4.1 million tons in the early 1990s. Special efforts have been made to promote extensive and intensive inland fish farming, modernize coastal fisheries, and encourage deep-sea fishing through joint ventures. These efforts led to a more than fourfold increase in coastal fish production from 520,000 tons in FY 1950 to 2.4 million tons in FY 1990. The increase in inland fish production was even more dramatic, increasing almost eightfold from 218,000 tons in FY 1950 to 1.7 million tons in FY 1990. The value of fish and processed fish exports increased from less than 1 percent of the total value of exports in FY 1960 to 3.6 percent in FY 1993.

Major Fishes

The important marine fish in the mid-1990s are mackerel, sardines, Bombay duck, shark, ray, percoids, croaker, carangid, sole, ribbonfish, whitebait, tuna, silverbelly, prawn, and cuttlefish. The main freshwater fish are carp and catfish; the main brackish-water fish are hilsa (a variety of shad), and mullet.

Economical Benefits

Great potential exists for expanding the nation's fishing industry. India's exclusive economic zone, stretching 200 nautical miles (370 km) into the Indian Ocean, encompasses more than 2 million square kilometers. In the mid-1980s, only about 33 percent of that area was being exploited. The potential annual catch from the area has been estimated at 4.5 million tons. In addition to this marine zone, India has about 14,000 km^2 of brackish water available for aquaculture, of which only 600 km^2 were being farmed in the early 1990s; about 16,000 km^2 of freshwater lakes, ponds, and swamps; and nearly 64,000 kilometers of rivers and streams.

In 1990, there were 1.7 million full-time fishermen, 1.3 million part-time fishermen, and 2.3 million occasional fishermen, many of whom worked as saltmakers, ferrymen, or seamen, or operated boats for hire. In the early 1990s, the fishing fleet consisted of 180,000 traditional craft powered by sails or oars, 26,000 motorized traditional craft, and some 34,000 mechanized boats.

Research and Training

Fisheries research and training institutions are supported by central and state governments that deserve much of the credit for the expansion and improvements in the Indian fishing industry. The principal fisheries research institutions, all of which operate under the Indian Council of Agricultural Research, are the Central Institute of Marine Fisheries Research at Kochi (formerly Cochin), Kerala; the Central Inland Fisheries Institute at Barrackpore, West Bengal; and the Central Institute of Fisheries Technology at Willingdon Island near Kochi. Most fishery training is provided by the Central Institute for Fishery Education in Mumbai, which has ancillary institutions in Barrackpore, Agra (Uttar Pradesh), and Hyderabad (Andhra Pradesh). The Central Fisheries Corporation in Calcutta is instrumental in bringing about improvements in fishing methods, ice production, processing, storing, marketing, and constructing and repairing fishing vessels. Operating under a 1972 law, the Marine Products Export Development Authority (MPEDA), headquartered in Kochi, has made several market surveys abroad and has been instrumental in introducing and enforcing hygiene standards that have gained for Indian fishery export products a reputation for cleanliness and quality.

Programmes

The implementation of two programs for inland fisheries—establishing fish farmers' development agencies and the National Programme of Fish Seed Development—has led to encouragingly increased production, which reached 1.5 million tons during FY 1990, up from 0.9 million tons in FY 1984. A network of 313 fish farmers' development agencies was functioning in 1992. Under the National Programme of Fish Seed Development, forty fish-seed hatcheries were commissioned. Fish-seed production doubled from 5 billion fry in FY 1983 to 10 billion fry in FY 1989. A new program using organic waste for aquaculture was started in FY 1986. Inland fish production as a percent of total fish production increased from 36 percent in FY 1980 to 40 percent by FY 1990.

Major Harbours

Apart from four main fishing harbours—Mangalore (Karnataka), Kochi (Kerala), Chennai (Tamil Nadu), Vishakhapatnam (Andhra Pradesh), and Roychowk in Kolkata (West Bengal)—twenty-three minor fishing harbours and ninety-five fish-landing centres are designated to provide landing and berthing facilities to fishing craft. The harbours at Vishakhapatnam, Kochi, and Roychowk were completed by 1980; the one at Madras was completed in the 1980s. A major fishing harbor was under construction at Sassoon Dock in Mumbai in the early 1990s, as were thirteen additional minor fishing harbours and eighteen small landing centres. By early 1990, there were 225 deep-sea fishing vessels operating in India's exclusive economic zone. Of these, 165 were owned by Indian shipping companies, and the rest were chartered foreign fishing vessels. The government provides subsidies to poor fishermen so that they can motorize their traditional craft to increase the range and frequency of operation, with a consequent increase in the catch and earnings. A total of about 26,171 traditional craft had been motorized under the program by 1992.

Restrictions

The banning of trawling by chartered foreign vessels and the speedy motorization of traditional fishing craft in the 1980s led to a quantum jump in marine fish production in the late 1980s. The export of marine products rose from 97,179 tons (Rs531 billion) in FY 1987 to 210,800 tons (Rs17.4 trillion) in FY 1992, making India one of the world's leading seafood exporting nations. This achievement was largely a result of significant advancements in India's freezing facilities since the 1960s, advancements that enabled India's seafood products to meet international standards. Frozen shrimp, a high-value item, has become the dominant seafood export. Other significant export items are frozen frog legs, frozen lobster tails, dried fish, and shark fins, much of which is exported to seafood-loving Japan. During the eighth plan, marine products were identified as having major export potential.

Institutes

There are several specialized institutes that train fishermen. The Central Institute of Fisheries Nautical and Engineering Training in Juhu instructs operators of deep-sea fishing vessels and technicians for shore establishments. It has facilities in Madras and Vishakhapatnam for about 500 trainees a year. The Integrated Fisheries Project, also headquartered in Kochi, was established for the processing, popularizing, and marketing of unusual fish. Another training organization, the Central Institute of Coastal Engineering for Fisheries in Bangalore, has done techno-economic feasibility studies on locations of fishing harbor sites and brackish-water fish farms. There are 14 Fisheries colleges functioning in various states of the country, providing Professional Fisheries education with a view of developing Professionalism in the field of Fisheries. The College Of Fisheries, Panangad, College of Fisheries, Mangalore etc.are some to name. To improve returns to fishermen and provide better products for consumers, several states have organized marketing cooperatives for fishermen. Nevertheless, most traditional fishermen rely on household members or local fish merchants for the disposal of their catches. In some places, marketing is carried on entirely by fisher women who carry small quantities in containers on their heads to nearby places. Good wholesale or retail markets are rare.

Fish Farming

Fish farming is the principal form of aquaculture, while other methods may fall under mariculture. Fish farming involves raising fish commercially in tanks or enclosures, usually for food. A facility that releases juvenile fish into the wild for recreational fishing or to supplement a species' natural numbers is generally referred to as a fish hatchery. The most common fish species raised by fish farms are, in order, salmon, carp, tilapia, catfish and cod.

Increasing demands on wild fisheries by commercial fishing has caused widespread overfishing. Fish farming offers an alternative solution to the increasing market demand for fish and fish protein.

Major Categories of Fish Farms

There are two kinds of aquaculture: extensive aquaculture based on local photosynthetical production and intensive aquaculture, in which the fish are fed with external food supply. The management of these two kinds of aquaculture systems are completely different.

Extensive Aquaculture

Limiting for growth here is the available food supply by natural sources, commonly zooplankton feeding on pelagic algae or benthic animals, such as crustaceans and mollusks. Tilapia species filter feed directly on phytoplankton, which makes higher production possible. The photosynthetic production can be increased by fertilizing the pond water with artificial fertilizer mixtures, such as potash, phosphorus, nitrogen and micro-elements. Because most fish are carnivorous, they occupy a higher place in the trophic chain and therefore only a tiny fraction of primary photosynthetic production (typically 1%) will be converted into harvest-able fish. As a result, without additional feeding the fish harvest will not exceed 200 kilograms of fish per hectare per year, equivalent to 1% of the gross photosynthetic production.

A second point of concern is the risk of algal blooms. When temperatures, nutrient supply and available sunlight are optimal for algal growth, algae multiply their biomass at an exponential rate, eventually leading to an exhaustion of available nutrients and a subsequent die-off. The decaying algal biomass will deplete the oxygen in the pond water because it blocks out the sun and pollutes it with organic and inorganic solutes (such as ammonium ions), which can (and frequently do) lead to massive loss of fish.

Another option is to use a wetland system such as that of Veta La Palma.

In order to tap all available food sources in the pond, the aquaculturist will choose fish species which occupy different places in the pond ecosystem, e.g., a filter algae feeder such as tilapia, a benthic feeder such as carp or catfish and a zooplankton feeder (various carps) or submerged weeds feeder such as grass carp.

Despite these limitations significant fish farming industries use these methods. In the Czech Republic thousands of natural and semi-natural ponds are harvested each year for trout and carp. The large ponds around Trebon were built from around 1650 and are still in use.

Sources:

- Introduction to Aquaculture, college notes, Department of Aquaculture, Wageningen University
- Aquaculture: training manual, second edition, Donald R. Swift, ISBN 0-85238-194-8.

Intensive Aquaculture

In these kinds of systems fish production per unit of surface can be increased at will, as long as sufficient oxygen, fresh water and food are provided. Because of the requirement of sufficient fresh water, a massive water purification system must be integrated in the fish farm. A clever way to achieve this is the combination of hydroponic horticulture and water treatment. The exception to this rule are cages which are placed in a river or sea, which supplements the fish crop with sufficient oxygenated water. Some environmentalists object to this practice.

The cost of inputs per unit of fish weight is higher than in extensive farming, especially because of the high cost of fish food, which must contain a much higher level of protein (up to 60%) than cattle food and a balanced amino acid composition as well. However, these higher protein level requirements are a consequence of the higher food conversion efficiency (FCR — kg of feed per kg of animal produced) of aquatic animals. Fish like salmon have FCR's in the 1.1 kg of feed per kg of salmon range whereas chickens are in the 2.5 kg of feed per kg of chicken range. Fish don't have to stand up or keep warm and this eliminates a lot of carbohydrates and fats in the diet, required to provide this energy. This frequently is offset by the lower land costs and the higher productions which can be obtained due to the high level of input control.

Essential here is aeration of the water, as fish need a sufficient oxygen level for growth. This is achieved by bubbling, cascade flow or aqueous oxygen. Catfish, Clarias ssp. can breathe atmospheric air and can tolerate much higher levels of pollutants than trout or salmon, which makes aeration and water purification less necessary and makes Clarias species especially suited for intensive fish production. In some Clarias farms about 10% of the water volume can consist of fish biomass.

The risk of infections by parasites like fish lice, fungi (Saprolegnia ssp.), intestinal worms (such as nematodes or trematodes), bacteria (e.g., Yersinia ssp, Pseudomonas ssp.), and protozoa (such as Dinoflagellates) is similar to animal husbandry, especially at high population densities. However, animal husbandry is a larger and more technologically mature area of human agriculture and better solutions to pathogen problem exist. Intensive aquaculture does have to provide adequate water quality (oxygen, ammonia, nitrite, etc.) levels to minimize stress, which makes the pathogen problem more difficult. This means, intensive aquaculture requires tight monitoring and a high level of expertise of the fish farmer.

Very high intensity recycle aquaculture systems (RAS), where there is control over all the production parameters, are being used for high value species. By recycling the water, very little water is used per unit of production. However, the process does have high capital and operating costs. The higher cost structures mean that RAS is only economical for high value products like broodstock for egg production, fingerlings for net pen aquaculture operations, sturgeon production, research animals and some special niche markets like live fish.

Raising ornamental cold water fish (goldfish or koi), although theoretically much more profitable due to the higher income per weight of fish produced, has never been successfully carried out until very recently. The increased incidences of dangerous viral diseases of koi Carp, together with the high value of the fish has led to initiatives in closed system koi breeding and growing in a number of countries. Today there are a few commercially successful intensive koi growing facilities in the UK, Germany and Israel.

Some producers have adapted their intensive systems in an effort to provide consumers with fish that do not carry dormant forms of viruses and diseases.

Specific Types of Fish Farms

Within intensive and extensive aquaculture methods there are numerous specific types of fish farms, each has benefits and applications unique to its design.

Integrated Recycling Systems

One of the largest problems with freshwater aquaculture is that it can use a million gallons of water per acre (about 1 m^3 of water per m^2) each year. Extended water purification systems allow for the reuse (recycling) of local water.

The largest-scale pure fish farms use a system derived (admittedly much refined) from the New Alchemy Institute in the 1970s. Basically, large plastic fish tanks are placed in a greenhouse. A hydroponic bed is placed near, above or between them. When tilapia are raised in the tanks, they are able to eat algae, which naturally grows in the tanks when the tanks are properly fertilized.

The tank water is slowly circulated to the hydroponic beds where the tilapia waste feeds a commercial plant crops. Carefully cultured microorganisms in the hydroponic bed convert ammonia to nitrates, and the plants are fertilized by the nitrates and phosphates. Other wastes are strained out by the hydroponic media, which doubles as an aerated pebble-bed filter.

This system, properly tuned, produces more edible protein per unit area than any other. A wide variety of plants can grow well in the hydroponic beds. Most growers concentrate on herbs (e.g. parsley and basil), which command premium prices in small quantities all year long. The most common customers are restaurant wholesalers.

Since the system lives in a greenhouse, it adapts to almost all temperate climates, and may also adapt to tropical climates. The main environmental impact is discharge of water that must be salted to maintain the fishes' electrolyte balance. Current growers use a variety of proprietary tricks to keep fish healthy, reducing their expenses for salt and waste water discharge permits. Some veterinary authorities speculate that ultraviolet ozone disinfectant systems (widely used for ornamental fish) may play a prominent part in keeping the Tilapia healthy with recirculated water.

A number of large, well-capitalized ventures in this area have failed. Managing both the biology and markets is complicated.

Irrigation Ditch or Pond Systems

These use irrigation ditches or farm ponds to raise fish. The basic requirement is to have a ditch or pond that retains water, possibly with an above-ground irrigation system (many irrigation systems use buried pipes with headers.) Using this method, one can store one's water allotment in ponds or ditches, usually lined with bentonite clay. In small systems the fish are often fed commercial fish food, and their waste products can help fertilize the fields. In larger ponds, the pond grows water plants and algae as fish food. Some of the most successful ponds grow introduced strains of plants, as well as introduced strains of fish.

Control of water quality is crucial. Fertilizing, clarifying and pH control of the water can increase yields substantially, as long as eutrophication is prevented and oxygen levels stay high. Yields can be low if the fish grow ill from electrolyte stress.

Composite Fish Culture

The Composite fish culture system is a technology developed in India by the Indian Council of Agricultural Research in the 1970s. In this system both local and imported fish species, a combination of five or six fish species is used in a single fish pond. These species are selected so that they do not compete for food among them having different types of food habitats. As a result the food available in all the parts of the pond is used.

Fish used in this system include catla and silver carp which are surface feeders, rohu a column feeder and mrigal and common carp which are bottom feeders. Other fish will also feed on the excreta of the common carp and this helps contribute to the efficiency of the system which in optimal conditions will produce 3000-6000 kg of fish per hectare per year.

Cage System

Fish cages are placed in lakes, bayous, ponds, rivers or oceans to contain and protect fish until they can be harvested. They can be constructed of a wide variety of components. Fish are stocked in cages, artificially fed, and harvested when they reach market size. A few advantages of fish farming with cages are that many types of waters can be used (rivers, lakes, filled quarries, etc.), many types of fish can be raised, and fish farming can co-exist with sport fishing and other water uses. Cage farming of fishes in open seas is also gaining popularity.

Concerns of disease, poaching, poor water quality, etc., lead some to believe that in general, pond systems are easier to manage and simpler to start. Also, past occurrences of cage-failures leading to escapes, have raised concern regarding the culture of non-native fish species in open-water cages. Even though the cage-industry has made numerous technological advances in cage construction in recent years, the concern for escapes remains valid.

Classic Fry Farming

Trout and other sport fish are often raised from eggs to fry or fingerlings and then trucked to streams and released. Normally, the fry are raised in long, shallow concrete tanks, fed with fresh stream water. The fry receive commercial fish food in pellets. While not as efficient as the New Alchemists' method, it is also far simpler, and has been used for many years to stock streams with sport fish. European eel (Anguilla anguilla) aquaculturalists procure a limited supply of glass eels, juvenile stages of the European eel which swim north from the Sargasso Sea breeding grounds, for their farms. The European eel is threatened with extinction because of the excessive catch of glass eels by Spanish fishermen and overfishing of adult eels in, e.g., the Dutch IJsselmeer, Netherlands. As per 2005, no one has managed to breed the European eel in captivity.

Criticisms

The issue of feeds in fish farming has been a controversial one. Many cultured fishes (tilapia, carp, catfish, many others) require no meat or fish products in their diets. Top-level carnivores (most salmon species) depend on fish feed of which a portion is usually derived from wild caught fish (anchovies, menhaden, etc.). Vegetable-derived proteins have successfully replaced fish meal in feeds for carnivorous fishes, but vegetable-derived oils have not successfully been incorporated into the diets of carnivores.

Secondly, farmed fish are kept in concentrations never seen in the wild (e.g. 50,000 fish in a 2-acre (8,100 m^2) area.) with each fish occupying less room than the average bathtub. This can cause several forms of pollution. Packed tightly, fish rub against each other and the sides of their cages, damaging their fins and tails and becoming sickened with various diseases and infections. This also causes stress.

However, fish tend also to be animals that aggregate into large schools at high density. Most successful aquaculture species are schooling species, which do not have social problems at high density. Aquaculturists tend to feel that operating a rearing system above its design capacity or above the social density limit of the fish will result in decreased growth rate and increased FCR (food conversion ratio-kg dry feed/kg of fish produced), which will result in increased cost and risk of health problems along with a decrease in profits. Stressing the animals is not desirable, but the concept of and measurement of stress must be viewed from the perspective of the animal using the scientific method.

Some species of sea lice have been noted to target farmed coho and Atlantic salmon. Such parasites have been shown to have an effect on nearby wild fish. One place that has garnered international media attention is British Columbia's Broughton Archipelago. There, juvenile wild salmon must "run a gauntlet" of large fish farms located off-shore near river outlets before making their way to sea. It is alleged that the farms cause such severe sea lice infestations that one study predicted a 99%

collapse in the wild salmon population in another four years. This claim, however, has been criticized by numerous scientists who question the correlation between increased fish farming and increases in sea lice infestation among wild salmon.

Because of parasite problems, some aquaculture operators frequently use strong antibiotic drugs to keep the fish alive (but many fish still die prematurely at rates of up to 30 percent). In some cases, these drugs have entered the environment. Additionally, the residual presence of these drugs in human food products has become controversial. Use of antibiotics in food production is thought to increase the prevalence of antibiotic resistance in human diseases. At some facilities, the use of antibiotic drugs in aquaculture has decreased considerably due to vaccinations and other techniques. However, most fish farming operations still use antibiotics, many of which escape into the surrounding environment.

The lice and pathogen problems of the 1990s facilitated the development of current treatment methods for sea lice and pathogens. These developments reduced the stress from parasite/pathogen problems. However, being in an ocean environment, the transfer of disease organisms from the wild fish to the aquaculture fish is an ever-present risk..

The very large number of fish kept long-term in a single location contributes to habitat destruction of the nearby areas. The high concentrations of fish produce a significant amount of condensed faeces, often contaminated with drugs, which again affect local waterways. However, these effects are very local to the actual fish farm site and are minimal to non-measurable in high current sites.

Other potential problems faced by aquaculturists are the obtaining of various permits and water-use rights, profitability, concerns about invasive species and genetic engineering depending on what species are involved, and interaction with the United Nations Convention on the Law of the Sea.

Indoor Fish Farming

An alternative to outdoor open ocean cage aquaculture, one in which the risk of environmental damage is high, is through the use of a recirculation aquaculture system (RAS). A RAS is a series of culture tanks and filters where water is continuously recycled and monitored to keep optimal conditions year round. To prevent the deterioration of water quality, the water is treated mechanically through the removal of particulate matter and biologically through the conversion of harmful accumulated chemicals into nontoxic ones.

Other treatments such as UV sterilization, ozonation, and oxygen injection are also used to maintain optimal water quality. Through this system, many of the environmental drawbacks of aquaculture are minimized including escaped fish, water usage, and the introduction of pollutants. The practices also increased feed-use efficiency growth by providing optimum water quality.

One of the drawbacks to recirculation aquaculture systems is water exchange. However, the rate of water exchange can be reduced through aquaponics, such as the incorporation of hydroponically grown plants and denitrification (Klas et al., 2006). Both methods reduce the amount of nitrate in the water, and can potentially eliminate the need for water exchanges, closing the aquaculture system from the environment. The amount of interaction between the aquaculture system and the environment can be measured through the cumulative feed burden (CFB kg/M3), which measures the amount of feed that goes into the RAS relative to the amount of water and waste discharged.

Because of its high capital and operating costs, RAS has generally been restricted to practices such as broodstock maturation, larval rearing, fingerling production, research animal production, SPF (specific pathogen free) animal production, and caviar and ornamental fish production. Although the use of RAS for other species is considered by many aquaculturalists to be impractical, there has been some limited successful implementation of this with high value product such as barramundi, sturgeon and live tilapia in the US.

7

Value Added Products from Fish

Fish Cutlet

One of the many products developed by the Central Institute of Fisheries Technology is fish cutlets. The basic raw material required for preparation of this product is cooked fish or 'fish 'kheema' (fish meat picked from whole fish by means of a meat picking machine).

- Cooked fish meat : 1000g
- Salt : 25g (approx-to taste)
- OIl : 125 ml
- Green chilli : 15 g
- Ginger : 25g
- Onion : 250 g
- Potato (cooked) : 500 g
- Pepper (powder) : 3 g (to taste)
- Clove (powered) : 3 g
- Cinnamon (powered) : 2g (to taste)
- Turmeric : 2 g
- Eggs : 4 Nos
- *Bread powder : 200 g.*

Method of preparation:

- Cook fish mince in boiling water for 20 min.
- Drain off the water. (In case of whole fish, dress the fish and cook for 30 min. and drain
- Remove skin, scales and bones and separate the meat

- Add salt and turmeric to the cooked meat and mix well
- Fry chopped onions in oil till brown. Fry chilli and ginger. Mix these with the cooked meat
- Add mashed potato and spices and mix well with the meat.
- Shape 40 g each of this in oval or round form, dip in beaten eggs, roll in bread powder store in deep freezer.
- Thaw and fry in oil before use.

Fish Balls

There are several varieties of fish which do not command a ready market as fresh fish, but are comparable to many table fish in nutritive value and other attributes. One of the ways of ensuring effective utilisation of such fish is to process ready-to-serve or ready-to-cook value added 'convenience' products, for which there already exists great demand from within the country as also from abroad. Fish ball is one such product prepared using fish mince and starch that can be processed as a coated product or as a heat-processed product in a suitable fluid medium.

Method of preparation:

- Mix mince prepared from fish using a mechanical meat bone separator after heading, gutting and washing thoroughly with 1 % salt and 5% corn starch (if required, spices like garlic, ginger etc. can be added).
- Prepare balls, 2-3 cm in diameter, from the resultant mass and cook in boiling 1% brine for 5-10 minutes.
- Cool the cooked balls after which they are battered and breaded.
- Pack the balls preferably in thermoformed trays as such or after flash frying in hot vegetable oil.
- Preserve by freezing and store at-18°C.

Though mince from different species of fish can be used for processing fish balls, threadfin bream (Nemipterus japonicus), Pallikora (Otolithus argenteusj and Barracuda (Sphyraena spp.) are found to yield satisfactory products. Mince from freshwater fishes such as Rohu (Labeo rohita) and Catla (Catla catla) also can be used. However, in such cases, the mince needs to be refined by passing through a mechanical strainer to get rid of the interstitial spines.

Fish Pickle

Traditionally, pickles made of vegetables like lime, gooseberry, ginger, garlic etc. are used as an important side dish along with meals. Though such pickles made of fish or meat were practically unknown in the past, such products have now become very popular and products under several brand names are now available in the market.

Ingredients:

- Fish (dressed and cut into small pieces : 1 kg
- Mustard : 10 g
- Green chilli (cut into pieces) : 50 g
- Garlic (peeled) : 200 g
- Ginger (peeled and chopped) : 150 g
- Chilli powder : 50 g
- Turmeric powder : 2 g
- Gingelly oil : 200 g
- Vinegar (acetic acid 1.5%) : 400 ml
- Salt : 60g
- Pepper (powdered) : 2.5g
- Sugar : 10g
- Cardamom, clove, cinnamon (powdered) : 1.5 g.

Method of Preparation

Mix the fish thoroughly with 3% of its weight of salt and keep for two hours. Light salted and partially dried fish also may be used. Fry the fish in minimum quantity of oil. Set apart the fried fish.

Fry the ingredients (mustard, green chillies, garlic, ginger) in the remaining quantity of oil and then add chilli powder, pepper powder and turmeric powder and mix well over low flame for a few minutes. Remove from fire, add fried fish and mix well. When cooled, add vinegar, powdered cardamom, clove, cinnamon, sugar and remaining salt and mix thoroughly. Sufficient quantity of boiled and cooled water may be added to cover the ingredients well. Transfer to clean, sterile g1ass bottles and seal with acid proof caps. Take care to see that there is a layer of oil over the contents in the bottle.

Flexible pouches made of 12 micron polyester laminated with l 18 micron LD-HD co-extruded film can also be use for packing the pickle.

Prawn Pickle

Ingredients:

- Prawn (peeled) : 1 kg
- Green chilli (small pieces) : 50 g
- Ginger (small pieces) : 150 g
- Garlic : 200 g
- Chilli powder : 35 g

- Turmeric powder : 2 g
- Gingelly oil : 200 ml
- Vinegar (1.5% acetic acid) (boiled and cooled) : 300 ml
- Salt (approx. 60 g) : to taste
- Sugar : 5 g.

Method of Preparation

Mix the peeled prawns with salt (3% on the basis of weight of prawns) and dry under sun for 1-2 hours. Fry the prawn in minimum quantity of oil and set apart. Fry garlic, ginger and green chilli in remaining quantity of oil. When brown in colour, add chilli powder and turmeric powder and mix together over a low flame. Remove from flame, add the prawns and mix thoroughly. Allow to cool and add vinegar, sugar and remaining salt. If necessary, add 1 % acetic acid to make up the consistency. Fill in clean, dry bottles taking care to provide a layer of oil at the top covering the contents.

Flexible pouches made of 12 μ polyester laminated with 118μ LD-HD co-extruded film can also be used for packing the pickle.

Fish Soup Powder

Soup powders prepared from different materials like vegetables, meat, egg, chicken etc. are very popular and widely consumed in different parts of the world. These are rich in dietary constituents like protein, vitamin, fat and minerals. However, soups making use of fish are not so common. But fishes of low economic value can very well be used for making soup powder.

Chief Ingredients:

Picked cooked meat	: 750 g
Salt	: 170 g
Fat	: 125 g
Onions	: 750 g
Coriander	: 12 g
Starch	: 250 g
Milk powder	: 100 g
Sugar	: 30 g
Pepper (powder)	: 15g
Ascorbic acid	: 1.5 g
Carboxymethyl cellulose	: 3 g
Monosodium glutamate	: 10 g.

Method of preparation:

- Blending of the processed fish meat: *Disperse the processed fish meat in about 150 ml of water and blend the material in a waring blender.*
- Incorporation of fried ingredients and reblending : *Fry the onion in vegetable fat in quantities as shown in the composition above. When the onion becomes light brown, add the powdered coriander and pepper and stir. Remove from fire. Add the fired onion and all other ingredients other than milk powder to the blended fish and again blend till it becomes a thick fine paste.*
- Drying : *Pour the whole mass in thin layer in aluminium trays and dry in artificial dryer at about 70° C.*
- Powdering and incorporation of milk powder: *Powder the dried mass, add the skimmed milk powder and powder well to get a homogenous product.*
- Packing : *Pack the soup powder in airtight, polythene lined, aluminium foil bags or in cans. (Storage life of the product is one year).*

Preparation for Consumption

Boil 5 g of the powder in 100 ml water for 5 minutes. The soup is now ready for the table.

Fish Wafers

Dried, ready-to-fry-and-serve wafers, using a carbohydrate as main base and incorporating salt and several other ingredients with or without spices are very popular in most parts of the country. Such products are known by different names in different languages as-Kondattarn' in Malayalam, 'Vathal' in Tamil, `Sandings' in Kanarese, `Odiyalu' in Telugu and `Tikiya' in Bengali. Recipe for such a product enriched with fish protein and method of its preparation is given below.

Ingredients:

- Processed fish meat : 2 kg
- Corn flour :1 kg
- Tapioca starch : 2 kg
- Common salt : 50 g
- Water : 3.5 litres.

Method of preparation:

- Homogenise the processed fish meat with 1 litre of water for 10 mins. in a mechanical grinding machine
- Add the cornflour, tapioca starch and salt and rest of the water and blend the whole mass for one hour

- Spread the homogenised mass uniformly in aluminium trays in a thin layer of 1-2 mm thickness and cook in steam for 3-5 mins.
- Cool to room temperature.
- Cut the cooked material into desired shapes and dry under sun or preferably in artificial dryer (at 45° C to 50° C) to a moisture content below 10%
- Pack suitable lots of the dried product in sealed polythene bags or glass bottles and store it in a cool and dry place till marketing.

The product can be stored in good condition for two years.

Permitted food colours can be incorporated, if needed, at the time of mixing the other ingredients with the processed fish meat in order to get desired colour.

Generally, this type of product is used as side dish after frying in oil.

Source : *Central Institute of Fisheries Technology, Cochin.*

Ready-to-serve Fish Curry in Flexible Pouches

Fish curry processed in flexible pouches is a product of great value these days. In the existing method, metal cans are used as containers.

Disadvantages of existing storage containers for ready-to-serve fish curry:

- Metal cans impart an undesirable taste to the product on storage
- Metal cans are not economical to use in India
- There is likelihood of leaks from containers if the metal containers are not strong
- Flexible pouches used are not heat stable and have various other disadvantages.

Improved method of packaging ready-to-serve fish curry:

- Central Institute of Fisheries Technology (CIFT) has developed a suitable three-layer configuration of flexible pouches for packaging ready-to-eat fish curry.
- This is a retortable flexible pouch based on Polyesters Aluminium foil/Cast polypropylene.
- These flexible pouches are manufactured in India employing the configuration developed by CIFT.
- CIFT has standardised the process for the production of fish curry in these pouches using over pressure autoclave. The curry processed in this way has remained without any change for over a year at room temperature.

Value added Products from Rainbow Trout

Rainbow trout is an exotic coldwater species which was introduced in many parts of India. The Himalayan foothills, Kashmir, upper streams of Western Ghats in Karnataka, Tamilnadu and Kerala are ideal for rainbow trout culture. In India, there is a high demand for this fish among the urban population. Presently, this fish is marketed in India in fresh chilled condition. This species is a ideal for the preparation of a variety of value added products.

Handling practices for fresh trout:

- The newly caught fish should be gutted, washed and packed in ice for better preservation.
- By adopting good chilling practices, fresh fish can also be preserved in whole condition
- To reduce bacterial contamination, fish should be gutted, gills removed and washed thoroughly before packing in ice. The fish can be marketed as whole or as fillets. The yield at different stages of processing is given below:

Stages of processing	***Yield***
Skin on fillets	42.1-48.1 %
Skin less fillets	39.6-42.5 %
Mince	33.9-38.7 %

The proximate composition of farmed rainbow trout is as follows:

Moisture	***78.10 %***
Crude Protein	19.80 %
Fat	0.62 %
Ash	0.61 %

Frozen Trout

Rainbow trout can be kept in good condition by freezing and holding in a cold store. Packing individual fish in polyethylene cover is effective in reducing dehydration and oxidation of the frozen product. Frozen whole trout remained in good condition at-18 ° C for 12 months.

Smoked Trout

The most common form of value addition of trout is as whole gutted, hot smoked product. It is cooked enough during the process, so as to be presented as a ready to eat product. CIFT has standardized the technique for the preparation of smoked products from this fish.

The whole fish was; beheaded, gutted, split open and cold blanched in 10% brine for 15 minutes. It was then hot smoked at 60°C for three hours to obtain the desired colour and flavour. Storage studies under chilled conditions showed that vacuum, packed smoked trout had a longer shelf life of 11 weeks when compared to air packed samples.

Coated Fillets

Coated product is a much sought after value added item. Battering and breading improve he appearance. Skinless fillets obtained, from whole trout were cold blanched in 5% salt and 0.1% citric acid for 10 minutes before breading and battering. This product remained good for twelve months in frozen storage at-18 ° C.

Canned Trout

The whole fish was gutted, washed and made into steaks of, uniform size. It was cold blanched in 10% NaCl containing 0.5% citric acid for 10 minutes and smoked at 45 ° C for two hours and given a partial drying for one hour. Approximately 120g of steaks were finger packed in 307 X 109 TFS cans. The cans were heat processed at 121°C in refined groundnut oil. Smoked and canned trout steak in oil is an excellent product with a shelf life of more than one year at ambient conditions

Wastewater Treatment through Aquaculture

With the increasing population in the country, the quantity of wastewaters generated has been increasing beyond the treatment capacities, apart from host of industrial effluents and solid wastes in recent years. Intense efforts are being made at treating the domestic sewage to make the effluents suitable for discharge into the natural waters.

Concept of biological treatment:

- Biological treatment involves systematic use of natural activity of the bacteria for biochemical reactions, resulting in oxidation of organic matter into CO_2, H_2O, N_2 and SO_4.
- The processes used widely for treatment of domestic sewage are activated sludge and trickling filter methods, oxidation/waste stabilization ponds, aerated lagoons and variations of anaerobic treatment systems;
- The latest one is the Upflow Anaerobic Sludge Blanket (UASB) process. The traditional practices of recycling sewage through agriculture, horticulture and aquaculture, being basically biological processes, have been in vogue in several countries. The sewage-fed fish culture in bheries of Calcutta is world-famous. The emphasis in these practices has been on the recovery of nutrients from the wastewater.
- Taking cue from these practices and deriving from the new databases in different disciplines of wastewater management, aquaculture is being proposed and standardized as a tool for treatment of domestic sewage.

Treatment Process

The wastewater treatment system through aquaculture includes sewage intake system, duckweed culture complex, sewage-fed fish-pond, depuration pond and outlet systems:

- The duckweed culture complex comprises a series of duckweed ponds where aquatic macrophytes like Spirodela, Wolffia and Lemna are grown. The wastewater is taken on gravity or pumped through intake system into duckweed culture system where it is retained for 2 days before allowing to fish-ponds.
- The model developed for treatment of 1 MLD wastewater consists of 18 duckweed ponds of 25 m x 8 m x 1 m, constructed in 3 rows, thus allowing water to pass through a series of three duckweed ponds into fish-ponds by gradient.
- The system includes 2 fish ponds of 50 m x 20 m x 2 m and two marketing ponds/depuration ponds of 40 m x 20 m x 2 m. the system receives primary treated sewage, after removal of solid materials.
- On the other hand, the treatment systems operated at two different locations of Bhubaneswar city for treatment of 8 MLD sewage, however, is designed with certain modifications for handling large volume of wastewater.
- As the optimum intake BOD levels for effective treatment are in the range of 100-150 mg/l, it may be necessary to incorporate an anaerobic unit where organic load and BOD levels are very high.
- The duckweed culture unit helps in removal of heavy metals and other chemical residues that otherwise get into human food chain through cultured fish. These also serve as nutrient pumps, reduce eutrophication effects and provide oxygen from their photosynthesis activity. The wastewater with BOD5 levels of about 100 mg/l can be treated with a total retention time of 5 days, with final BOD5 levels brought down to 15-20 mg/l, meeting the required standards of different parameters for discharge into natural waters.
- Further, taking advantage of high productivity and carrying capacity in sewage-fed system, production level of 3-4 tonnes of carps/ha is achieved in fish culture ponds. The system provides for a biological treatment system with high potential of resource recovery in terms of duckweed and fish. The main limitation of this biological treatment system, however, is the reduction of efficiency of treatment during the winter period and the regions with temperate climatic condition. Comparatively low requirement of land of about 1.0 ha for treatment of 1 MLD sewage and resource earnings partially meeting working costs, makes the system an ideal one for treating wastewater before discharge into natural waters.

Trout Farming in Himachal Pradesh

Zone II and III of Himachal Pradesh have vast potentials for the culture of highly prized fish "Rainbow Trout". The agroclimatic conditions of the area under these two zones are very congenial for cold water aquaculture. Recent findings indicate that trout can be cultured at lower elevation even upto 1000 m MSL, provided the optimum water quality and quantity is ensured.

Selection of Site

For trout culture, such a site should be selected where the quality and quantity of water is available through perennial source like spring, river, etc.

Construction of Ponds

The cemented ponds/raceways are required for the culture of trout fish. The rectangular tanks are better than the circular cisterns. The economical size of a trout race way should be 12-15 m x 2-3 ms x1.2-.5 m with an inlet and an outlet for overflow of water fixed with wire mesh screws to prevent the exit of stocked species. There should be a drain pipe at the bottom of the pond to facilitate the harvesting as well as the cleaning of tank from time to time.

Water Supply in the Farm

The water supply in trout farm should be through a filter bed/sedimentation tank. There is a lot of silt problem in this area particularly during monsoon season when the water is turbid which is not good for trout farming. The quantity of water required for a trout farm is related with the stocking density, size of fish as well as the water temperature.

Therefore, it is necessary to regulate the flow of water very carefully. For example, 30,000 fries need water flow of about 15 L/minute, the fish below 250 g needs a water flow of 0.5 L/kg/minute at 10-12°C. The water flow of above mentioned economical size of tank should be 52 m3/hour for stocking of fingerlings of 5-50 g at 15°C. Thus, the water flow is regulated in such a way that fishes should not assemble at one place and also do not move fast. The flow of water should be increased with the increase of water temperature.

Physico-chemical Parameters Required for a Trout Farm

The physico-chemical parameters responsible for the successful culture of trout are temperature, dissolved oxygen, pH and turbidity.

Temperature

The fish thrives well within the temperature range of 5 to 18°C, but it has been found to tolerate the water temperature upto 25°C in the region without any mortality. However, the maximum growth is obtained within the temperature range of 10 to 18°C.

Dissolved Oxygen

The oxygen concentration range is from 5.8 to 9.5 mg/L. If the oxygen concentration is 5 mg/L, it is better to increase the flow of water.

pH

A neutral or slightly alkaline pH is best for the trout. The tolerable minimum and maximum pH values are 4.5 and 9.2, respectively. However, pH range is ideal for the growth of this fish.

Turbidity

The crystal clear water is required and there should not be any contamination. The turbidity should not be more than 25 cm.

Stocking Density

It is related with the water supply, water temperature, quality/water and types of feed. If water temperature is above 20°C, the stocking density should be less than the recommended density. The fry fingerlings (5 to 50 g) is stocked at the rate of 20 kg fish per cubic meter of water surface area.

Supply of Feed

The quantity of feed mainly depends on the water temperature and size of fish. If the water temperature is above 18°C, the recommended feed should be reduced to just half of the required amount and above 20°C, better to stop the feeding. The feeding should also be suspended on a cloudy day and when the water is turbid.

A practical feed formula is given below :

Ingredients	*Percentage of ingredients*	*Quantity for preparing 10 kg of feed(kg)*
Fish meal	50	5
Soyaflakes	10	1
Groundnut cake	20	2
Wheat flour	10	1
Linseed oil	9	0.9
Supplevit-M	1	0.1
Choline chloride	0.1	0.01

Feeding @ 4-6 % is necessary for the fingerlings for better growth but due consideration should also be given to the water temperature for following the feeding schedule. At the water temperature range of 10-12°C, feeding schedule of 6% is optimum but when it increases to 15°C, the feeding schedule to be lowered to 4% and beyond 19°C, it should be just 50% of the optimum schedule. The optimum growth rate per month is 80 g.

Table Size Fish

The fish after gaining the weight of 250 g is advisable to be harvested because beyond this size the growth is slow and rearing is uneconomical.

Hygiene

The cleanliness is a very important factor in trout farming. The trout should be cleaned and disinfected either with 10% formaline or 4 ppm KMno4 solution periodically. The infected fish should be immediately removed from the tank and due care should be taken to consult some fishery expert regarding the disease, if any.

Ornamental Fish Farming

Ornamental fish keeping and its propagation has been an interesting activity for many, which provide not only aesthetic pleasure but also financial openings. About 600 ornamental fish species have been reported worldwide from various aquatic environments. Indian waters possess a rich diversity of ornamental fish, with over 100 indigenous varieties, in addition to a similar number of exotic species that are bred in captivity

Fish Species/Types Suitable for Breeding

Among the indigenous and exotic freshwater species, the varieties having good demand can be bred and reared for commercial purpose. Species that can be produced easily and popular as commercial varieties are coming under egg-layers and livebearers.

Livebearer species:

- *Guppies (*Peocillia reticulata*)*
- *Moly (*Mollinesia *sp.)*
- *Sword tail (*Xiphophorus *sp.)*
- Platy.

Egg-layers:

- *Goldfish (*Carassius auratus*)*
- *Koi carp (*Cyprinus carpio *var.* koi*)*
- Zebra danio (Brachydanio rerio)
- *Black window tetra (*Cymnocro-cymbus *sp.)*
- *Neon tetra (*Hyphesso-brycon innesi*)*
- *Serpae tetra (*Hyphessobrycon callistus*).*

Others;

- Bubbles-nest builders
- *Angelfish (*Pterophyllum scalare*)*
- *Red-line torpedo fish (*Puntius denisoni*)*
- *Loaches (*Botia *sp.)*

- *Leaf-fish (*Nandus nandus.
- *Snakehead (*Channa orientalis*).*

A beginner should start the work on breeding of any live-bearer followed by goldfish or any other egg-layer for getting acquainted with the procedures on handling and maintenance of brood fish and young one. Good knowledge on the biology, feeding behaviour and ambient condition of the fish are prerequisites for breeding. Live food like Tubifex worms, Moina, earthworms' etc. for brood-stock and larval stages need special attention. The larvae similarly need infusoria, artemia naupli, plantons such as rotifers and smaller daphnia during the early stage. A unit for continuous production of live-food is, therefore necessary for successful maintenance of the unit. In most cases breeding is easy, but larval rearing might require special care. As a supplementary feeding the farmer can prepare on-site pelletted feed by using local agro-produce. To avoid health related problems, a proper water quality needs to be ensured by installing biofilters. The ornamental fishes can be bred at varied periods of the year.

Some Tips for the Successful Production of Ornamental Fishes

Breeding and rearing unit should be made near a constant supply of water and electricity. If the unit is located near the streams, it will be excellent where the unit can receive portable water and the rearing unit can be made flow-through.

Constant availability of agro-based by-products like oil cakes, rice polish and wheat bran, and animal-based protein such as fish meal and prawn-head meal, will facilitate preparation of pelleted diet for the fish. The brood-stock selected for breeding should be of superior quality so that they produce quality fish for sale. It is advisable that the young ones could be grown till their maturity. It not only gives enough experience of handling the fish but also helps in controlled selection.

Breeding and rearing unit may be established preferably nearer to airport/ railway station so that live fish could be easily transported to internal market as also for export. A fish breeder could concentrate preferably on a single marketable species to streamline management measures.

Proper knowledge on market demand, customer preference and overall, the operation of a marketing network through personal contacts and public relation is desirable.Pioneer and expert groups in this field may always be kept in touch with to keep track of recent developments in marketing as well as research advancements, through training.

Fresh Water Prawn (Indian River Prawn) Culture

About Fresh Water Prawn

Fresh Water Prawn (Macrobrachium malcolmsonii), the second largest fast-growing prawn occurs commonly in Indian rivers, draining into Bay of Bengal. Under monoculture systems production levels of 750-1,500 kg prawns/ha/8 months are achieved. Further, it is a compatible species for polyculture along with Indian major carps and Chinese carps, which may yield 400 kg prawns and 3000 kg carps/ha/yr. Since the seed requirement for the commercial farming of this species is not met from the natural resources, large-scale seed production under controlled conditions for year-round supply is extremely important. The technologies of large-scale seed production and grow-out culture have led to increased attention of the farmers and entrepreneurs for diversification of their culture practice.

Broodstock Management

Broodstock and berried females are essential component for continuous operation for seed production. The gonadal maturation of the species differs greatly in nature depending on the agro-climatic conditions. In the Ganga, the Hooghly and the Mahanadi river systems, the maturation and breeding start from May and continue till the end of October, whereas in the Godavari, the Krishna and the Cauvery systems it commences from April and continue till November. Under pond conditions, sexual maturity generally occur after attaining a maximum size of 60-70 mm. Berried females are recorded year-round in some of the ponds. The ratio of berried females in total population is found to be higher during August-September and during this period they carry good quantity of eggs (8000-80,000). Prawns breed 3-4 times in a season. Successful community breeding and year-round seed production under captive conditions is possible employing air-lift bio-filter re-circulatory system.

Spawning and Larval Rearing

Mating takes place immediately after pre-mating moult in matured female and spawning occurs few hours after mating. Incubation period of eggs lasts between 10-15 days depending upon the optimum water temperature of 28-30°C. However, at lower temperature, the incubation period is prolonged to more than 21 days. Hatching of fully developed 1st zoea takes place through the body stretching of the zoea, which breaks the eggshell and comes out from the egg and starts swimming as plankton.

Different larval rearing technologies viz., static, flow-through, clear or green water, closed or semi-closed, with or without circulation systems of larval rearing of prawn species under hatchery conditions have been developed with varying degrees of success. The green water technique has been claimed to increase the post-larval production by 10-20% over other techniques. But higher mortalities are generally encountered due to rise in pH and uncontrolled algal bloom. Further, increase in numbers of adult Artemia, due to abundance of feed in green water, contributes to accumulation of ammonia in the culture medium. The production of post-larvae (PL) in large numbers may be possible following airlift bio-filter re-circulatory system. The larvae passed through 11 zoeal stages before attaining PL within a period of 39-60 days at salinity and temperature ranging from 18-20‰ and 28-31°C, respectively, with the production of 10-20 PL/l.

Bio-filter equipped with air-lift re-circulation has shown promising results in maintaining favourable water quality in different rearing media with enhanced rate of post-larval production. Among various physico-chemical parameters of rearing medium, temperature, pH, dissolved oxygen, total hardness, total alkalinity, salinity and ammonical nitrogen have been registered as the potent parameters which influence greatly the growth, metamorphosis and survival of larvae in captive rearing. The respective parameters within the ranges of 28-30°C, 7.8-8.2, 3000-4500 ppm, 80-150 ppm, 18-20‰ and 0.02-0.12 ppm are considered optimum during the larval rearing of the species.

Larval Feeding

Various feed items viz., Artemia nauplii, minute zooplankton, especially cladocerans, copepods, rotifers, flesh of prawn and fish, molluscan meat, earthworms, tubificid worm, egg custard, cut pieces of goat/hen viscera etc., are used during larval rearing. Among these Artemia nauplii have been proved as an excellent larval food for the prawn larvae. At the beginning, freshly hatched Artemia nauplii are provided to the 1st stage zoea at 1 g/30,000 larvae twice daily up to 15 days or till they attain stage VI. Thereafter, the feed is given once daily along with egg custard and mussel meat/tubificid worm four times daily.

Harvesting of Post-larvae

Harvesting of post-larvae of prawn is rather difficult due to their crawling habit. Therefore, both turn-down and drain siphoning of water are commonly used for harvesting. But due to longer duration for attaining post-larval stage the above methods are neither useful nor safe. Further, the presence of post-larvae in the larval tank affects the growth and survival of advanced larvae due to competition for food and cannibalism. Hence, the need for an ideal device for regular harvest of post-larvae from the rearing unit is very much essential. String shell is therefore devised and is successfully used for phase wise harvest of post-larvae during larval rearing. Post-larval survival and production rates, following air-left bio-filter re-circulatory system, are in the range of 10-20 PL/l.

Post-larval Rearing

Optimum growth, production and survival of prawns can be achieved in grow-out ponds on stocking the nursery reared juveniles rather than stocking directly with the freshly metamorphosed post-larvae. Post-larvae slowly adopt themselves to freshwater. Optimum growth and survival of healthy juveniles during post-larval rearing is achieved at salinity of 10‰. Post-larval rearing can be done both in well-prepared earthen ponds with adequate aeration facility and inside the hatchery following bio-filter re-circulatory system. Stocking density, feed and water quality management play the major role in raising healthy juveniles during rearing. Stocking density between 10-15 post-larvae/l is ideal. Among various feed items, egg custard along with chopped freshwater mussel meat have been found to be more effective in maintaining good growth. Water quality parameters, viz., water temperature, pH, dissolved oxygen and dissolved ammonia in the ranges of 27.5-30°C, 7.8-8.3, 4.4-5.2 ppm and 0.02-0.03 ppm, respectively are considered to be optimum.

Grow-out Culture

Grow-out system of prawn is similar to that of freshwater fish. As the prawns can migrate from one pond to other due to its crawling habit, it is necessary to have the pond embankment 0.5 m higher from the water level. Sandy-clay pond bottom is considered to be favourable for better growth. Undrainable ponds may be treated with conventional piscicides for eradication of predatory and weed fishes. Stocking density of 30,000 to 50,000/ha is advocated for semi-intensive farming. Ponds with the facility of water exchange and aeration can be used for intensive farming where stocking density may be increased to 1 lakh/ha. Temperature is the most important factor which directly controls the growth and survival of prawns. Temperatures above 35°C or below 14°C are generally reported to be lethal and 29-31°C is optimal.

Males grow faster than females. Mixture of groundnut oil cake and fish meal in the proportion 1:1 is used as supplementary feed. A production of 500-1000 kg/ha in six months of rearing are achieved under monoculture with the stocking density of 30,000-50,000. In polyculture, M. malcolmsonii at a stocking density of 10,000-20,000/ha along with carps at density of 2,500-3,500 nos/ha, a production of 300-400 kg prawn and 2000-3000 kg carps can also be raised.

8

Shrimp Farm Management

Shrimp are swimming, decapod crustaceans classified in the infraorder Caridea, found widely around the world in both fresh and salt water. Adult shrimp are filter feeding benthic animals living close to the bottom. They can live in schools and can swim rapidly backwards. Shrimp are an important food source for larger animals from fish to whales. They have a high tolerance to toxins in polluted areas, and may contribute to high toxin levels in their predators. Together with prawns, shrimp are widely caught and farmed for human consumption.

Taxonomy

A number of more or less unrelated crustaceans share the word "shrimp" in their common name. Examples are the mantis shrimp and the opossum or mysid shrimp, both of which belong to the same class (Malacostraca) as decapods including shrimp, but constitute two different orders within it, the Stomatopoda and the Mysidacea. Triops longicaudatus and Triops cancriformis are also popular animals in freshwater aquaria, and are often called shrimp, although they belong instead to the Notostraca, a quite unrelated group.

Biological Definition

Shrimp (Caridea) typically have two pairs of claws, and the second segment of the abdomen overlaps the segments on either side. The abdomen shows a pronounced caridean bend.

Prawns (Dendrobanchiata), such as this Penaeus species, typically have three pairs of claws, and even-sized segments on the abdomen.

The class Malacostraca contains about half of the crustaceans. The members of this class have a primitive body plan that can be described as shrimp-like, consisting of a 5-8-7 body plan. They have a small carapace that encloses the head and the thorax, and have a muscular abdomen for swimming. They also have a thin exoskeleton to maintain a light weight. These general characters are common in all members of the class.

The class can be further divided into the decapods, which are even still divided into the dendrobranchiates (prawns) and the carideans (shrimp and snapping shrimp) .

The prawns have sequentially overlapping body segments (segment one covers the segment two, segment two covers segment three, etc), chelate (claw like) first three leg pairs, and have a very basic larval body type.

The shrimps also have overlapping segments, however, in a different pattern (segment two overlaps segments one and three), only the first two leg pairs are chelate, and they have a more complex larval form.

Biologists distinguish the true shrimp from the true prawn because of the differences in their gill structures. The gill structure is lamellar in shrimp but branching in prawns. The easiest practical way to separate true shrimps from true prawns is to examine the second abdominal segment. The second segment of a shrimp overlaps both the first and the third segment, while the second segment of a prawn overlaps only the third segment .

Commercial and Culinary Definition

While in biological terms shrimps and prawns belong to different suborders of Decapoda, they are very similar in appearance. In commercial farming and fisheries, the terms shrimp and prawn are often used interchangeably. However, recent aquaculture literature increasingly uses the term "prawn" only for the freshwater forms of palaemonids and "shrimp" for the marine penaeids .

In the United Kingdom, the word "prawn" is more common on menus than "shrimp"; while the opposite is the case in North America. The term "prawn" is also loosely used to describe any large shrimp, especially those that come 15 (or fewer) to the pound (such as "king prawns", yet sometimes known as "jumbo shrimp"). Australia and some other Commonwealth nations follow this British usage to an even greater extent, using the word "prawn" almost exclusively. When Australian comedian Paul Hogan used the phrase, "I'll slip an extra shrimp on the barbie for you" in an American television advertisement , it was intended to make what he was saying easier for his American audience to understand, and was thus a deliberate distortion of what an Australian would typically say.

In Britain very small crustaceans with a brownish shell are called shrimp, and are used to make potted shrimp. They are also used in dishes where they are not the primary ingredient.

Consumption

As with other seafood, shrimp is high in calcium, iodine and protein but low in food energy. A shrimp-based meal is also a significant source of cholesterol, from

122 mg to 251 mg per 100 g of shrimp, depending on the method of preparation Shrimp consumption, however, is considered healthy for the circulatory system because the lack of significant levels of saturated fat in shrimp means that the high cholesterol content in shrimp actually improves the ratio of LDL to HDL cholesterol and lowers triglycerides .

Shrimp and other shellfish are among the most common food allergens . They are not kosher and thus are forbidden in Jewish cuisine. However, according to some Madh'hab, shrimp are halal, and therefore are permissible in Islamic cuisine.

Commercial Fishing

Common commercial methods for catching shrimp and prawns include otter trawls, cast nets, seines, shrimp baiting and dip netting. Trawling involves the use of a system of nets. In some parts of the Pacific Northwest, fishing with baited traps is also common.

The following table shows the yearly weight of shrimp and prawns captured globally in millions of tonnes .

Production	*1999*	*2000*	*2001*	*2002*	*2003*	*2004*	*2005*
Million tonnes	3.03	3.09	2.96	2.97	3.55	3.54	3.42

The highest rates of incidental catch of non-target species is associated with shrimp trawling. In 1997, the FAO documented the estimated bycatch and discard levels from shrimp fisheries around the world. They found discard rates as high as 20 pounds for every pound of shrimp, with a world average of 5.7 pounds for every pound of shrimp .

Trawl nets in general, and shrimp trawls in particular, have been identified as sources of mortality for species of finfish and cetaceans . Bycatch is often discarded dead or dying by the time it is returned to the sea, and may alter the ecological balance in discarded regions . Worldwide, shrimp trawl fisheries generate about 2% of the world's catch of fish in weight, but result in more than one third of the global bycatch total.

Shrimp Farm

A shrimp farm is an aquaculture business that cultivates marine shrimp or prawns for human consumption. Commercial shrimp farming began in the 1970s, and production grew steeply, particularly to service the U.S., Japan and Western Europe. Global production of farmed shrimp reached more than 1.6 million tonnes in 2003, representing a value of nearly 9,000 million U.S. dollars. About 75% of farmed shrimp in produced in Asia, in particular in China and Thailand. The other 25% comes mainly from Latin America, where Brazil is the largest producer. Thailand is the largest exporting nation.

Shrimp farming has grown from a traditional, small-scale businesses in Southeast Asia into a global industry. Technological advances have led to growing shrimp at ever higher densities. Broodstock is shipped worldwide. Virtually all farmed shrimp are penaeids (i.e., of the family Penaeidae), and just two species of shrimp—the Penaeus vannamei (Pacific white shrimp) and the Penaeus monodon (giant tiger prawn)—account for roughly 80% of all farmed shrimp. These industrial monocultures are very susceptible to diseases, which have caused several regional wipe-outs of farm shrimp populations. Increasing ecological problems, repeated disease outbreaks, and pressure and criticism from both NGOs and consumer countries led to changes in the industry in the late 1990s and generally stronger regulation by governments. In 1999, a program aimed at developing and promoting more sustainable farming practices was initiated, including governmental bodies, industry representatives, and environmental organizations.

History and Geography

Indonesians and others have farmed shrimp for centuries, using traditional low-density methods. Indonesian brackish water ponds called tambaks can be traced back as far as the 15th century. They used small scale ponds for monoculture or polycultured with other species such as milkfish, or in rotation with rice, using the rice paddies for shrimp cultures during the dry season, when no rice could be grown. Such cultures often were in coastal areas or on river banks. Mangrove areas were favoured because of their abundant natural shrimp. Wild juvenile shrimp were trapped in ponds and reared on naturally occurring organisms in the water until they reached the desired size for harvesting.

Industrial shrimp farming can be traced to the 1930s, when Japanese agrarians spawned and cultivated Kuruma shrimp (Penaeus japonicus) for the first time. By the 1960s, a small industry had developed in Japan. Commercial shrimp farming began to grow rapidly in the late 1960s and early 1970s. Technological advances led to more intensive forms of farming, and growing market demand led to worldwide proliferation of shrimp farms, concentrated in tropical and sub-tropical regions. Growing consumer demand in the early 1980s coincided with faltering wild catches, creating a booming industry. Taiwan was an early adopter and a major producer in the 1980s; its production collapsed beginning in 1988 due to poor management practices and disease. In Thailand, large-scale production expanded rapidly from 1985. In South America, Ecuador pioneered shrimp farming, where it expanded dramatically from 1978. Brazil had been active in shrimp farming since 1974, but trade boomed there only in the 1990s, making the country a major producer within a few years. Today, there are marine shrimp farms in over fifty countries.

Farming Methods

When shrimp farming emerged to satisfy demand that had surpassed the wild fishery's capacity, the subsistence farming methods of old were rapidly replaced by the more productive practices required to serve a global market. Industrial farming at first followed traditional methods with so-called "extensive" farms, compensating for low density with increased pond sizes: instead of ponds of just a few hectares, ponds of sizes up to Template: Convert/were used and huge areas of mangroves were cleared in some areas. Technological advances made more intensive practices possible that increase yield per area, helping reduce pressure to convert more land. Semi-intensive and intensive farms appeared, where the shrimp were reared on artificial feeds and ponds were actively managed. Although many extensive farms remain, new farms typically are of the (semi-)intensive kind. Until the mid-1980s, most farms were stocked with young wild animals, called postlarvae, typically caught locally. Postlarvae fishing became an important economic sector in many countries. To counteract the depletion of fishing grounds and to ensure a steady supply of young shrimp, the industry started breeding shrimp in hatcheries.

Life Cycle

Shrimp mature and breed only in a marine habitat. The females lay 50,000 to 1 million eggs, which hatch after some 24 hours into tiny nauplii. These nauplii feed on yolk reserves within their body and then metamorphose into zoeae. This second larval stage feeds in the wild on algae and after a few days morph again into myses. The myses look akin to tiny shrimp and feed on algae and zooplankton. After another three to four days they metamorphose a final time into postlarvae: young shrimp that have adult characteristics. The whole process takes about 12 days from hatching. In the wild, postlarvae then migrate into estuaries, which are rich in nutrients and low in salinity. They migrate back into open waters when they mature. Adult shrimp are benthic animals living primarily on the sea bottom.

Supply Chain

In shrimp farming, this lifecycle occurs under controlled conditions. The reasons to do so include more intensive farming, improved size control resulting in more uniformly sized shrimp, and better predator control, but also the ability to accelerate growth and maturation by controlling the climate (especially in farms in the temperate zones, using greenhouses). There are three different stages:

- Hatcheries breed shrimp and produce nauplii or even postlarvae, which they sell to farms. Large shrimp farms maintain their own hatcheries and sell nauplii or postlarvae to smaller farms in the region.
- Nurseries grow postlarvae and accustom them to the marine conditions in the growout ponds.
- In the growout ponds the shrimp are grown from juveniles to marketable size, which takes between three to six months.

Most farms produce one to two harvests a year; in tropical climates, even three are possible. Because of the need for salt water, shrimp farms are located on or near a coast. Inland shrimp farms have also been tried in some regions, but the need to ship salt water and competition for land with agricultural users led to problems. Thailand banned inland shrimp farms in 1999.

Hatcheries

Small-scale hatcheries are very common throughout Southeast Asia. Often run as family businesses and using a low-technology approach, they use small tanks (less than ten tons) and often low animal densities. They are susceptible to disease, but due to their small size, they can typically restart production quickly after disinfection. The survival rate is anywhere between zero and 90%, depending on a wide range of factors, including disease, the weather, and the experience of the operator.

Greenwater hatcheries are medium-sized hatcheries using large tanks with low animal densities. To feed the shrimp larvae, an algal bloom is induced in the tanks. The survival rate is about 40%.

Galveston hatcheries (named after Galveston, Texas, where they were developed) are large-scale, industrial hatcheries using a closed and tightly controlled environment. They breed the shrimp at high densities in large (15 to 30 ton) tanks. Survival rates vary between zero and 80%, but typically achieve 50%.

In hatcheries, the developing shrimp are fed on a diet of algae and later also brine shrimp nauplii, sometimes (especially in industrial hatcheries) augmented by artificial diets. The diet of later stages also includes fresh or freeze-dried animal protein, for example krill. Nutrition and medication (such as antibiotics) fed to the brine shrimp nauplii are passed on to the shrimp that eat them.

Nurseries

Many farms have nurseries where the postlarval shrimp are grown into juveniles for another three weeks in separate ponds, tanks, or so-called raceways. A raceway is a rectangular, long, shallow tank through which water flows continuously.

In a typical nursery, there are 150 to 200 animals per square metre. They are fed on a high-protein diet for at most about three weeks before they are moved to the growout ponds. At that time, they weigh between one and two grams. The water salinity is adjusted gradually to that of the growout ponds.

Farmers refer to postlarvae as "PLs", with the number of days suffixed (i.e., PL-1, PL-2, etc.). They are ready to be transferred to the growout ponds after their gills have branched, which occurs around PL-13 to PL-17 (about 25 days after hatching). Nursing is not absolutely necessary, but is favoured by many farms because it makes for better food utilization, improves the size uniformity, helps utilize the infrastructure better, and can be done in a controlled environment to increase the harvest. The main disadvantage of nurseries is that some of the postlarval shrimp die upon the transfer to the growout pond.

Some farms do not use a nursery but stock the postlarvae directly in the growout ponds after having acclimated them to the appropriate temperature and salinity levels in an acclimation tank. Over the course of a few days, the water in these tanks is changed gradually to match that of the growout ponds. The animal density should not exceed 500/liter for young postlarvae and 50/liter for larger ones, such as PL-15.

Growout

In the growout phase, the shrimp are grown to maturity. The postlarvae are transferred to ponds where they are fed until they reach marketable size, which takes about another three to six months. Harvesting the shrimp is done by fishing them from the ponds using nets or by draining the ponds. Pond sizes and the level of technical infrastructure vary.

Extensive shrimp farms using traditional low-density methods are invariably located on a coast and often in mangrove areas. The ponds range from just a few to more than 100 hectares; shrimp are stocked at low densities (2–3 animals per square metre, or 25,000/ha). The tides provide for some water exchange, and the shrimp feed on naturally occurring organisms. In some areas, farmers even grow wild shrimp by just opening the gates and impounding wild larvae. Prevalent in poorer or less developed countries where land prices are low, extensive farms produce annual yields from 50 to 500 kg/ha of shrimp (head-on weight). They have low production costs (US$1–3/kg live shrimp), are not very labour intensive, and do not require advanced technical skills.

Semi-intensive farms do not rely on tides for water exchange but use pumps and a planned pond layout. They can therefore be built above the high tide line. Pond sizes range from 2 to 30 ha; the stocking densities range from 10 to 30/m^2 (100,000–300,000/ha). At such densities, artificial feeding using industrially prepared shrimp feeds and fertilizing the pond to stimulate the growth of naturally occurring organisms become a necessity. Annual yields range from 500 to 5,000 kg/ha, while production costs are in the range of US$2–6/kg live shrimp. With densities above 15 animals per square metre, aeration is often required to prevent oxygen depletion. Productivity varies depending upon water temperature, thus it is common to have larger sized shrimp in some seasons than in others. Intensive farms use even smaller ponds (0.1–1.5 ha) and even higher stocking densities. The ponds are actively managed: they are aerated, there is a high water exchange to remove waste products and maintain water quality, and the shrimp are fed on specially designed diets, typically in the form of formulated pellets. Such farms produce annual yields between 5,000 and 20,000 kg/ha; a few super-intensive farms can produce as much as 100,000 kg/ha. They require an advanced technical infrastructure and highly trained professionals for constant monitoring of water quality and other pond conditions; their production costs are in the range of US$4–8/kg live shrimp.

Estimates on the production characteristics of shrimp farms vary. Most studies agree that about 55–60% of all shrimp farms worldwide are extensive farms, another 25–30% are semi-intensive, the rest being intensive farms. Regional variation is high, though, and reports wide discrepancies in the percentages claimed for individual countries by different studies.

Feeding the Shrimps

While extensive farms mainly rely on the natural productivity of the ponds, more intensively managed farms rely on artificial shrimp feeds, either exclusively or as a supplement to the organisms that naturally occur in a pond. A food chain is established in the ponds, based on the growth of phytoplankton. Fertilizers and mineral conditioners are used to boost the growth of the phytoplankton to accelerate the growth of the shrimps. Waste from the artificial food pellets and excrements of the shrimps can lead to the eutrophication of the ponds.

Artificial feeds come in the form of specially formulated, granulated pellets that disintegrate quickly. Up to 70% of such pellets are wasted, as they decay before the shrimps have eaten them. The shrimps are fed two to five times daily; the feeding can be done manually either from ashore or from boats, or using mechanized feeders distributed all over a pond. The feed conversion rate (FCR), i.e. the amount of food needed to produce a unit (e.g. one kilogram) of shrimp, is claimed by the industry to be around 1.2–2 in modern farms, but this is an optimum value that is not always attained in practice. For a farm to be profitable, a feed conversion rate below 2.5 is necessary; in older farms or under suboptimal pond conditions, the ratio may easily rise to 4:1. Lower FCRs result in a higher profit for the farm.

Farmed Species

Although there are many species of shrimp and prawn, only a few of the larger ones are actually cultivated, all of which belong to the family of penaeids (family Penaeidae), and within it to the genus Penaeus. Many species are unsuitable for farming: they are too small to be profitable, or simply stop growing when crowded together, or are too susceptible to diseases. The two species dominating the market are:

- Pacific white shrimp (Litopenaeus vannamei, also called "whiteleg shrimp") is the main species cultivated in western countries. Native to the Pacific coast from Mexico to Peru, it grows to a size of 23 cm. P. vannamei accounts for 95% of the production in Latin America. It is easy to breed in captivity, but succumbs to the Taura disease.
- Giant tiger prawn (P. monodon, also known as "black tiger shrimp") occurs in the wild in the Indian Ocean and in the Pacific Ocean from Japan to Australia. The largest of all the cultivated shrimp, it can grow to a length of 36 cm and is farmed in Asia. Because of its susceptibility to whitespot

disease and the difficulty of breeding it in captivity, it is gradually being replaced by L. vannamei since 2001.

Together, these two species account for about 80% of the whole farmed shrimp production. Other species being bred are:

- Western blue shrimp (P. stylirostris) was a popular choice for shrimp farming in the western hemisphere, until the IHHN virus wiped out nearly the whole population in the late 1980s. A few stocks survived and became resistant against this virus. When it was discovered that some of these were also resistant against the Taura virus, some farms again bred P. stylirostris from 1997 on.
- Chinese white shrimp (P. chinensis, also known as the fleshy prawn) occurs along the coast of China and the western coast of Korea and is being farmed in China. It grows to a maximum length of only 18 cm, but tolerates colder water (min. 16 °C). Once a major factor on the world market, it is today used almost exclusively for the Chinese domestic market after a disease wiped out nearly all the stocks in 1993.
- Kuruma shrimp (P. japonicus) is farmed primarily in Japan and Taiwan, but also in Australia; the only market is in Japan, where live Kuruma shrimp reach prices of the order of US$100 per pound ($220/kg).
- Indian white shrimp (P. indicus) is a native of the coasts of the Indian Ocean and is widely bred in India, Iran and the Middle East and along the African shores.
- Banana shrimp (P. merguiensis) is another cultured species from the coastal waters of the Indian Ocean, from Oman to Indonesia and Australia. It can be grown at high densities.

Several other species of Penaeus play only a very minor role in shrimp farming. Some other kinds of shrimp also can be farmed, e.g. the "Akiami paste shrimp" or Metapenaeus spp. Their total production from aquaculture is of the order of only about 25,000 tonnes per year, small in comparison to that of the penaeids.

Diseases

There are a variety of lethal viral diseases that affect shrimp. In the densely populated, monocultural farms such virus infections spread rapidly and may wipe out whole shrimp populations. A major transfer vector of many of these viruses is the water itself; and thus any virus outbreak also carries the danger of decimating shrimp living in the wild.

Yellowhead disease, called Hua leung in Thai, affects P. monodon throughout Southeast Asia. It had been reported first in Thailand in 1990. The disease is highly contagious and leads to mass mortality within 2 to 4 days. The cephalothorax of an infected shrimp turns yellow after a period of unusually high feeding activity ending abruptly, and the then moribund shrimp congregate near the surface of their pond before dying.

Whitespot syndrome is a disease caused by a family of related viruses. First reported in 1993 from Japanese P. japonicus cultures, it spread throughout Asia and then to the Americas. It has a wide host range and is highly lethal, leading to mortality rates of 100% within days. Symptoms include white spots on the carapace and a red hepatopancreas. Infected shrimp become lethargic before they die.

Taura syndrome was first reported from shrimp farms on the Taura river in Ecuador in 1992. The host of the virus causing the disease is P. vannamei, one of the two most commonly farmed shrimp. The disease spread rapidly, mainly through the shipping of infected animals and broodstock. Originally confined to farms in the Americas, it has also been propagated to Asian shrimp farms with the introduction of P. vannamei there. Birds are thought to be a route of infection between farms within one region.

Infectious Hypodermal and Hematopoietic Necrosis (IHHN) is a disease that causes mass mortality among P. stylirostris (as high as 90%) and severe deformations in P. vannamei. It occurs in Pacific farmed and wild shrimp, but not in wild shrimp on the Atlantic coast of the Americas.

There are also a number of bacterial infections that are lethal to shrimp. The most common is vibriosis, caused by the bacterium Vibrio spp. The shrimp become weak and disoriented and may have dark wounds on the cuticle. The mortality rate can exceed 70%. Another bacterial disease is Necrotising hepatopancreatitis (NHP); symptoms include a soft exoskeleton and fouling. Most such bacterial infections are strongly correlated to stressful conditions such as overcrowded ponds, high temperatures, and poor water quality: factors that positively influence the growth of bacteria. Treatment is done using antibiotics. Importing countries have repeatedly placed import bans on shrimp containing various antibiotics. One such antibiotic is chloramphenicol, which has been banned in the European Union since 1994, but continues to pose problems.

With their high mortality rates, diseases represent a very real danger to shrimp farmers, who may lose their income for the whole year if their ponds are infected. Since most diseases cannot yet be treated effectively, the industry's efforts are focused on preventing diseases to break out in the first place. Active water quality management helps avoid poor pond conditions favourable to the spread of diseases, and instead of using larvae from wild catches, specific pathogen free broodstocks raised in captivity in isolated environments and certified not to carry diseases are used increasingly. To avoid introducing diseases into such disease-free populations on a farm, there is also a trend to create more controlled environments in the ponds of (semi-)intensive farms, such as by lining them with plastic to avoid soil contact, and by minimizing water exchange in the ponds.

Economy

The total global production of farmed shrimp reached more than 1.6 million tonnes in 2003, representing a farm-gate value of nearly 9 billion U.S. dollars. This accounts for 25% of the total shrimp production that year (farming and wild catches combined). The largest single market for shrimp is the United States, importing between 500 – 600,000 tonnes of shrimp products yearly in the years 2003-2009. About 200,000 tonnes yearly are imported by Japan, while the European Union imported in 2006 another ca. 500,000 tonnes of tropical shrimps, with the largest importers being Spain and France. The EU also is a major importer or coldwater shrimp from catches, mainly Common shrimp (Crangon crangon) and Pandalidae such as Pandalus borealis; in 2006, these imports accounted for another ca. 200,000 tonnes.

The import prices for shrimp fluctuate wildly. In 2003 the import price per kilogram shrimp in the United States was US$ 8.80, slightly higher than in Japan at US$8.–. The average import price in the EU was only about US$5.–/kg; this much lower value is explained by the fact that the EU imports more coldwater shrimp (from catches) that are much smaller than the farmed warm water species and thus attain lower prices. In addition, Mediterranean Europe prefers head-on shrimp which weigh approximately 30% more but have a lower unit price.

About 75% of the world production of farmed shrimp comes from Asian countries; the two leading nations being China and Thailand, closely followed by Vietnam, Indonesia, and India. The other 25% are produced in the western hemisphere, where the South-American countries (Brazil, Ecuador, Mexico) dominate. In terms of export, Thailand is by far the leading nation with a market share of more than 30%, followed by China, Indonesia, and India, accounting each for about 10%. Other major export nations are Vietnam, Bangladesh, and Ecuador. Thailand exports nearly all of its production, while China uses most of its shrimp in the domestic market.

Disease problems have repeatedly impacted the shrimp production negatively. Besides the near-wipeout of P. chinensis in 1993, there were outbreaks of viral diseases that led to marked declines in the per-country production in 1996/97 in Thailand and repeatedly in Ecuador. In Ecuador alone, production suffered heavily in 1989 (IHHN), 1993 (Taura), and 1999 (whitespot). Another reason for sometimes wild changes in shrimp farm output are the import regulations of the destination countries, which do not allow shrimp contaminated by chemicals or antibiotics to be imported.

In the 1980s and through much of the 1990s, shrimp farming promised high profits. The investments required for extensive farms were low, especially in regions with low land prices and wages. For many tropical countries, especially those with poorer economies, shrimp farming was an attractive business, offering jobs and incomes for poor coastal populations and has, due to the high market prices of shrimp, provided many developing countries with non-negligible foreign currency earnings. Many shrimp farms were funded initially by the World Bank or substantially subsidized by local governments.

In the late 1990s, the economic situation changed. Governments and farmers alike were under increasing pressure from NGOs and the consumer countries, who criticized the practices of the trade. International trade conflicts erupted, such as import bans by consumer countries on shrimp containing antibiotics, the United States' shrimp import ban against Thailand in 2004 as a measure against Thai shrimp fishers not using Turtle Excluder Devices in their nets, or the "anti-dumping" case initiated by U.S. shrimp fishers in 2002 against shrimp farmers worldwide, which resulted two years later in the U.S. imposing anti-dumping tariffs of the order of about 10% against many producer countries (except China, which received a 112% duty). Diseases caused significant economic losses. In Ecuador, where shrimp farming was a major export sector (the other two are Bananas and Oil), the whitespot outbreak of 1999 caused an estimated 130,000 workers to lose their jobs. Furthermore, shrimp prices dropped sharply in 2000. All of these factors contributed to the slowly growing acceptance by farmers that improved farming practices were needed, and resulted in tighter government regulation of the business, both of which internalized some of the external costs that were ignored during the boom years.

Socio-economic Aspects

Shrimp farming offers significant employment opportunities, which may help alleviate the poverty of the local coastal populations in many areas, if it is properly managed. The published literature on that topic shows large discrepancies, and much of the available data is of anecdotal nature. Estimates of the labour-intensiveness of shrimp farms range from about three times less to three times more than when the same area was used for rice paddies, with much regional variation and depending on the type of farms surveyed. In general, intensive shrimp farming requires more labour per unit area than extensive farming. Extensive farms cover much more land area and are often but not always located in areas where no agricultural land uses are possible. Supporting industries such as feed production or storage, handling, and trade companies should also not be neglected, even if not all of them are exclusive to shrimp farming.

Typically, workers on a shrimp farm can get better wages than with other employments. A global estimate from one study is that a shrimp farm worker can earn 1.5 – 3 times as much as in other jobs; a study from India arrived at a salary increase of about 1.6, and a report from Mexico states that the lowest paid job at shrimp farms was paid in 1996 at 1.22 times the average worker salary in the country.

NGOs have frequently criticized that most of the profits went to large conglomerates instead of to the local population. While this may be true in certain regions such as Ecuador, where most shrimp farms are owned by large companies, it does not apply in all cases. For instance in Thailand, most farms are owned by small local entrepreneurs, although there is a trend to vertically integrate the industries related to shrimp farming from feed producers to food processors and trade companies. A 1994 study reported that a farmer in Thailand could increase his income by a factor of ten by switching from growing rice to farming shrimp. An Indian study from 2003 arrives at similar figures for shrimp farming in the East Godavari district in Andhra Pradesh.

Whether the local population benefits from shrimp farming is also dependent on the availability of sufficiently trained people. Extensive farms tend to offer mainly seasonal jobs during harvest that do not require much training. In Ecuador, many of these positions are known to have been filled by migrant workers. More intensive farms have a need for year-round labour in more sophisticated jobs.

Ecological Impacts

Shrimp farms of all types, from extensive to super-intensive, can cause severe ecological problems wherever they are located. For extensive farms, huge areas of mangroves were cleared, reducing biodiversity. During the 1980s and 1990s, about 35% of the world's mangrove forests have vanished. Shrimp farming was a major cause of this, accounting for over a third of it according to one study; other studies report between 5% and 10% globally, with enormous regional variability. Other causes of mangrove destruction are population pressure, logging, pollution from other industries, or conversion to other uses such as salt pans. Mangroves, through their roots, help stabilize a coastline and capture sediments; their removal has led to a marked increase of erosion and less protection against floods. Mangrove estuaries are also especially rich and productive ecosystems and provide the spawning grounds for many species of fish, including many commercially important ones. Many countries have protected their mangroves and forbidden the construction of new shrimp farms in tidal or mangrove areas. The enforcement of the respective laws is often problematic, though, and especially in the least developed countries such as Bangladesh, Myanmar, or Vietnam the conversion of mangroves to shrimp farms remains an issue.

Intensive farms, while reducing the direct impact on the mangroves, have other problems. Their nutrient-rich effluents (industrial shrimp feeds disintegrate quickly, as little as 30% are actually eaten by the shrimp with a corresponding economic loss to the farmer, the rest is wasted) are typically discharged into the environment, seriously upsetting the ecological balance. These waste waters contain significant amounts of chemical fertilizers, pesticides, and antibiotics that cause pollution of the environment. Furthermore, releasing antibiotics in such ways injects them into the food chain and increases the risks of bacteria becoming resistant against them. However, most aquatic bacteria, unlike bacteria associated with terrestrial animals, are not zoonotic. Only a few disease transfers from animals to humans have been reported. Prolonged use of a pond can lead to an incremental build-up of a sludge at the pond's bottom from waste products and excrements. The sludge can be removed mechanically or dried and plowed to allow bio-decomposition, at least in areas without acid problems. Flushing a pond never completely removes this sludge, and eventually, the pond is abandoned, leaving behind a wasteland with the soil made unusable for any other purposes due to the high levels of salinity, acidity, and toxic chemicals.

A typical pond in an extensive farm can be used only a few years. An Indian study estimated the time to rehabilitate such lands to about 30 years. Thailand has banned inland shrimp farms since 1999 because they caused too much destruction of agricultural lands due to salination. A Thai study estimated that 60% of the shrimp farming area in Thailand was abandoned in the years 1989 – 1996. Much of these problems stem from using mangrove land that has high natural pyrite content (acid sulfate soil) and poor drainage. The shift to semi-intensive farming requires higher elevations for drain harvesting and low sulfide (pyrite) content to prevent acid formation when the soils shift from anaerobic to aerobic conditions.

The global nature of the shrimp farming business and in particular the shipment of broodstock and hatchery products throughout the world have not only introduced various shrimp species as exotic species, but also distributed the diseases the shrimp may carry worldwide. As a consequence, most broodstock shipment require health certificates and/or to be SPF (specific pathogen free) status. Many organizations lobby actively for consumers to avoid buying farmed shrimp; some also advocate the development of more sustainable farming methods.

A joint programme of the World Bank, the Network of Aquaculture Centres in Asia-Pacific (NACA), the WWF, and the FAO was established in August 1999 to study and propose improved practices for shrimp farming. Some existing attempts at sustainable export-oriented shrimp farming marketing the shrimp as "ecologically produced" are criticized by NGOs as being dishonest and trivial window-dressing.

Yet the industry has been slowly changing since about 1999. It has adopted the "best management practices" developed by e.g. the World Bank et al. programme and instituted educational programmes to promote them. Due to the mangrove protection laws enacted in many countries, new farms are usually of the (semi-)intensive kind, which are best constructed outside of mangrove areas anyway. There is a trend to create even more tightly controlled environments in these farms with the hope to achieve better disease prevention. Waste water treatment has attracted considerable attention; modern shrimp farms routinely have effluent treatment ponds where sediments are allowed to settle at the bottom and other residuals are filtered. As such improvements are costly, the World Bank et al. programme also recommends low-intensity polyculture farming for some areas. Since it has been discovered that mangrove soils are effective in filtering waste waters and tolerate high nitrate levels, the industry has also developed an interest in mangrove reforestation, although its contributions in that area are still minor. The long-term effects of these recommendations and industry trends cannot be evaluated conclusively yet.

Social Changes

Shrimp farming in many cases has far-reaching effects on the local coastal population. Especially in the boom years of the 1980s and 1990s, when the business was largely unregulated in many countries, the very fast expansion of the industry caused significant changes that sometimes were detrimental to the local population. Conflicts can be traced back to two root causes: competition for common resources such as land and water, and changes induced by wealth redistribution.

A significant problem causing much conflict in some regions, for instance in Bangladesh, are the land use rights. With shrimp farming, a new industry expanded into coastal areas and started to make exclusive use of previously public resources. In some areas, the rapid expansion resulted in the local coastal population being denied access to the coast by a continuous strip of shrimp farms with serious impacts on the local fisheries.

Such problems were compounded by poor ecological practices that caused a degradation of common resources (such as excessive use of freshwater to control the salinity of the ponds, causing the water table to sink and leading to the salination of freshwater aquifers by an inflow of salt water). With growing experience, countries usually introduced stronger governmental regulations and have taken steps to mitigate such problems, for instance through land zoning legislations. Some late adopters have even managed to avoid some problems through proactive legislation, e.g. Mexico. The situation in Mexico is unique owing to the strongly government-regulated market. Even after the liberalisation in the early 1990s, most shrimp farms are still owned and controlled by locals or local co-ops (ejidos).

Social tensions have occurred due to changes in the wealth distribution within populations. The effects of this are mixed, though, and the problems are not unique to shrimp farming. Changes in the distribution of wealth tend to induce changes in the power structure within a community. In some cases, there is a widening gap between the general population and local élites who have easier access to credits, subsidies, and permits and thus are more likely to become shrimp farmers and benefit more. In Bangladesh, on the other hand, local élites were opposing shrimp farming, which was controlled largely by an urban élite. Land concentrations in a few hands has been recognized to carry an increased risk of social and economic problems developing, especially if the landowners are non-local.

In general, it has been found that shrimp farming is accepted best and introduced most easily and with the greatest benefits for the local communities if the farms are owned by local people instead of by restricted remote élites or large companies because local owners have a direct interest in maintaining the environment and good relations with their neighbors, and because it avoids the formation of large-scale land property.

Freshwater prawn farming shares many characteristics and problems with marine shrimp farming. Unique problems are introduced by the main species' (the giant river prawn, Macrobrachium rosenbergii) developmental life cycle.

The global annual production of freshwater prawns (excluding crayfish and crabs) in 2003 was about 280,000 tonnes, of which China produced some 180,000 tonnes, followed by India and Thailand with some 35,000 tonnes each. China also produced about 370,000 tonnes of Chinese mitten crab (Eriocheir sinensis).

Shrimp Marketing

Shrimp are marketed and commercialized with several issues in mind. Most shrimp are sold frozen and marketed based on their categorization of presentation, grading, colour, and uniformity.

Presentation

The main forms of presentation are head on shell on (HOSO), shell on (SO), peeled tail on (PTO), peeled undeveined (PUD), peeled and deveined (P&D), and butterfly tail on (BTTY-TO). Sometimes a letter 'F' is placed in front of these abbreviation for the presentation in order to state that the shrimp comes from a farm (example: FSO-farm, shell on). European and Asian markets prefer the HOSO presentation (which is a whole shrimp), while the American shrimp market prefers the remaining presentations.

Grading

Shrimp are graded according to their count per weight. HOSO shrimps are graded in units per kilogram (30/40, 40/50, 50/60, etc. pcs/kg). The standard pack is in a 2 kg box, 10 boxes into a master carton. The remaining presentations are graded in units per pound (U15, 16/20, 21/25, 26/30, 31/35, 36/40, 41/50, etc pcs/lb). The standard pack is in a 5 lb box, 10 boxes into a master carton.

The numbers in the grading code indicate maximum and minimum quantity of pieces per unit weight, with U standing for "under".

Colour

HOSO shrimp are also graded according to their colour, with A1 being the lightest colour and A5 the darkest colour. Shrimp tend to take on the colour of their habitat, so sandy ponds tend to yield an A1 colour shrimp, and shrimp ponds with plastic black liners tend to yield A5 colour shrimp. A2-A3 colour shrimp are preferred for fresh commercialization, and A4-A5 colour shrimp are preferred for cooked commercialization. The other presentations are not usually graded by colour.

Uniformity

Another concept to grade is the uniformity. This is measured by visually selecting the ten largest and the ten smallest pieces from 1 kg of product. The two groups are then weighed separately and the weight of the large pieces group is divided by the weight of the small pieces group. The result is the uniformity factor. Normally the maximum accepted value for the uniformity factor is 1.5.

Preparation

Preparing shrimp for consumption usually involves removing the head, shell, tail, and "sand vein".

To de-shell a shrimp, the tail is held while gently removing the shell around the body. The tail can be detached completely at this point, or left attached for presentation purposes.

Removing the "vein" (a euphemism for the digestive tract) can be referred to as "deveining", though in fact shrimp do not have any real veins; they have an open circulatory system. The "vein" can be removed by making a shallow cut lengthwise down the outer curve of the shrimp's body, allowing the dark ribbon-like digestive tract to be removed with a pointed utensil. Alternatively, if the tail has been detached, the vein can be pinched at the tail end and pulled out completely with the fingers. The shrimp is then rinsed under cold running water.

Shrimp and prawns are versatile ingredients, and are often used as an accompaniment to fried rice. Common methods of preparation include baking, boiling, and frying.

Recipes using shrimp form part of the cuisine of many cultures. Strictly speaking, dishes containing scampi should be made from the Norway lobster, a shrimp-like crustacean more closely related to the lobster than shrimp, but in some places it is quite common for large shrimp to be used instead.

Wet shrimp is commonly used as a flavouring and as a soup base in Asian cuisines (such as Thai tom yum goong) while fried shrimp is popular in North America. In Europe, shrimp is very popular, forming a necessary ingredient in Spanish paella de marisco, Italian cacciucco, Portuguese caldeirada and many other seafood dishes. Shrimp curry is very popular in South Asia and Southeast Asia. Shrimp are also found in Latin and Caribbean dishes such as enchiladas and coconut shrimp. Other recipes include jambalaya, okonomiyaki, poon choi and bagoong. Shrimp are also consumed as salad, by frying, with rice, and as shrimp guvec (a dish baked in a clay pot) in the Western and Southern coasts of Turkey.

Life Cycle

Most shrimp mature and breed only in a marine habitat, although there are a small number of freshwater species. The females lay 50,000 to 1 million eggs, which hatch after some 24 hours into tiny nauplii. These nauplii feed on yolk reserves within their body and then undergo a metamorphosis into zoeae. This second larval stage feeds in the wild on algae and after a few days metamorphoses again into the third stage to become myses.

At this stage the myses already begin to appear like tiny versions of fully-developed adults and feed on algae and zooplankton. After another three to four days they metamorphose a final time into postlarvae: young shrimp having all the characteristics of adults.

The whole process takes about 12 days from hatching. In the wild, the marine postlarvae then migrate into estuaries, which are rich in nutrients and low in salinity. There they grow and eventually migrate back into open waters when they mature. Most adult shrimp are benthic animals living primarily on the sea floor.

Common shrimp species include pink, brown and snapping shrimp. Depending on the species and location, they grow from about 1.2 to 30 centimetres (0.47 to 12 in) long, and live between one and 6.5 years .

Fossil Record

The fossil record of shrimp is sparse, with only 57 exclusively fossil species known . The earliest of these cannot be assigned to any family, but date from the Lower Jurassic and Cretaceous .

Home Aquaria

Several types of shrimp are kept in home aquaria. Some are purely ornamental, while others are useful in controlling algae and removing debris . Freshwater shrimp commonly available for aquaria include the Japanese marsh shrimp (Caridina multidentata, also called "Amano shrimp," as their use in aquaria was pioneered by Takashi Amano), cherry shrimp (Neocaridina heteropoda), and ghost or glass shrimp (Palaemonetes spp.). Popular saltwater shrimp include the cleaner shrimp Lysmata amboinensis, the fire shrimp (Lysmata debelius) and the harlequin shrimp (Hymenocera picta).

Etymology

The term shrimp originated around the 14th century with the Middle English shrimpe, akin to the Middle Low German schrempen, and meaning to contract or wrinkle; and the Old Norse skorpna, meaning to shrivel up

Mud Crab

Mud crab is highly popular due to its great demand in the export market. The commercial scale mud crab culture is developing fast along the coastal areas of Andhra Pradesh, Tamil Nadu, Kerala and Karnataka.

Types of Mud Crabs

Mud crabs in the genus Scylla inhabit coastal areas, estuaries and backwaters.

i. Larger species :
   - The larger species is locally known as 'green mud crab'.
   - It grows to a maximum size of 22 cm carapace width and 2 kg in weight.
   - These are free living and distinguished by the polygonal markings present on all appendages.

ii. Smaller species :
   - The smaller species is known as 'red claw'.
   - This grows to a maximum size of 12.7 cm carapace width and 1.2 kg in weight.
   - It is without polygonal markings and has a burrowing habit.

Both species have good demand in the domestic as well as in the foreign market.

Culture Methods

i. Grow-out culture.
   - In this method, young crabs are grown for a period of 5 to 6 months till they attain desirable size.
   - Mud crab grow-out systems are generally pond based, with or without mangroves.
   - The pond size varies between 0.5-2 ha, with proper bunds and tidal water exchange.
   - Fencing is advisable if the pond is small. In large ponds where natural conditions are prevailing, strengthening is necessary along the outlet area.
   - Wild collected juvenile crabs of 10-100 g size are used for stocking.
   - The duration of the culture varies between 3-6 months.
   - The stocking rates are commonly between 1-3 crabs/m^2, with supplementary feeding.
   - Feeding is usually with trash fish (wet weight feeding rate-5% per day of the biomass), along with other locally available items.
   - Regular sampling is necessary to monitor the growth and general health, and to adjust the feeding rate.
   - Partial harvesting of marketable sized crabs can be started from 3rd month onwards. This 'stock-thinning' provides chances for better survival by reducing the mutual attacks and cannibalism.

ii. Fattening.

Soft shelled crabs are reared for a period of a few weeks till their exoskeleton gets hardened. These 'hard' crabs are locally known as "mud" (meat) and fetch three to four times better price than the soft crabs.

Fattening in Ponds:

- Fattening can be done in small tidal ponds between 0.025-0.2 ha with a water depth of 1 to 1.5 m.
- Before stocking the soft crabs in the pond, the bottom is prepared by draining the pond water, sun-drying and adding sufficient quantity of lime.
- Care is taken in strengthening the pond bunds without any holes and crevices. Special care is taken in the sluice area as these crabs have a tendency to escape through nearby areas of sluice gate. The inlet areas should be reinforced with bamboo matting inside the bund.
- The ponds are fenced properly using bamboo poles and nets along the sides of the bund, which incline towards the pond to prevent the escape of crabs.
- Soft crabs collected from local fishermen/crab merchants are stocked in the ponds preferably in the morning hours @ 0.5-2 crabs/m^2 according to the size of the crabs.
- For crabs of 550g and above, market demand is more. Hence it is ideal to stock

crabs belonging to this size group. In this case, the stocking density should not be more than 1 rab/m^2.

- Depending upon the location and availability of water crabs, 6-8 cycles of "fattening" can be carried out in a pond by repetitive stocking and harvesting.
- If the culture pond is big, it is better to divide the pond into different compartments of suitable sizes for stocking crabs of uniform sizes in the same compartment. This is good for manipulating the feeding and is easy to monitor and harvest.
- When the interval between stockings is wide, crabs of similar sizes can be stocked in one compartment.
- Sex-wise stocking in compartments is advantageous to reduce the attacks from more aggressive male crabs. It's better to provide shelters like old tyres, bamboo baskets, tiles etc. to minimize mutual attacks and cannibalism.

Fattening in pens and cages:

- Fattening also can be carried out in pens, floating net cages or bamboo cages in shallow estuarine waterways and inside large shrimp ponds with good tidal water influx.
- HDPE, netlon or bamboo splits can be used as the netting material.

- The size of the cage shall be preferably 3 m x 2 m x 1 m.
- The cages have to be arranged in a row so that feeding and monitoring can be easily carried out.
- A stocking of 10 crab/m^2 in cages and 5 crabs/m^2 for pens is recommended. Since the stocking rate is higher in cages, and to minimize the mutual attacks, the tips of the chelae can be removed while stocking.
- However, these methods are not commercialized like 'fattening' in ponds.

Between these two methods, Fattening is more advantageous since the culture period is short and profitable, when enough stocking material is assured. In India, grow-out culture is not popular, mainly due to non-availability of crab seeds and commercial feed.

Feeding

Crabs are daily fed with trash fish, brackish water clams or boiled chicken waste @ 5-8% of their body weight. If the feed is given twice a day, major share must be given during evening hours.

Water quality

The water quality parameters shall be maintained within the range shown below:

Salinity	15-25%
Temperature	26-30° C
Oxygen	> 3 ppm
pH	7.8-8.5

Harvesting and marketing:

- Periodically crabs are to be checked for their hardening.
- Harvesting should be done in the early morning hours or evening hours.
- The harvested crabs should be washed in good brackish water to remove the dirt and mud and carefully tied without breaking its legs.
- The harvested crabs have to be kept in moist conditions. It has to be kept away from sunlight, which has a negative effect on survival.

Economics of mud crab fattening-(6 crops/annum), (0.1 ha tidal pond)

A. Annual fixed cost	*Rs*
Pond (lease amount)	10,000
Sluice gate	5,000
Pond preparation, fencing and misc. charges	10,000

B. Operational cost (single crop)	
1. Cost of water crabs (400 crabs @ Rs. 120/kg)	36,000
2. Feed cost	10,000
3. Labour charges	3,000
	Sub total for one crop 49,000
	Total for 6 crops 2,94,000
C. Annual total cost	***3,19,000***

D. Yield and revenue	
Production of crabs per cycle	240 kg
Total revenue for 6 cycles (Rs.320/Kg)	4,60,800
E. Net profit	1,41,800

- Economics is given for a pond of suitable size which can be easily managed by small/marginal farmers. They can even go for smaller sizes.
- Stocking density is low (0.4 no./m2), since the suggested stocking size of the crab is about 750g.
- Feeding rate is 10% of the total biomass for the first week and 5% during rest of the period. It is better to use feeding trays to avoid wastage of feed and to maintain good water quality
- In well managed ponds, 8 'fattening' cycles can be carried out with 80-85% survival (here only 6 cycles with 75% survival are taken in to consideration).

9

Integrated Livestock-fish Farming

Fish-Duck Integrated Farming in Assam

Integrated fish farming has received attention in recent years in the North East India. Among the different livestock based system, fish duck integration is one of the most popular farming practices among woman fishers.

Advantages of Fish Duck Integration:

- Duck dropping act as feed and fertilizer for cultured fish in the pond.
- Duck collect 50% of their food naturally from the pond.
- Ducks keep the water body clean and increase dissolved oxygen by doubling action.
- Duck house can be constructed at the embankment or over the water surface, hence no need of extra place for it.
- Left over feed of duck used as supplementary feed for fish.
- Production of duck eggs, meat, fish and horticulture from the same unit area.

Selection of fish species:

- In this integration, fish species selection is one of the important point, where plankton feeder should be 60% and omnivorous should be 40%.
- A combination of six species viz, Catla (20%), Silver carp (20%), Rohu (20%), Mrigal (15%), Grass carp (10%) and Common carp (15%) should be stocked at density of 8000 to 8500 fingerlings/ha for the targeted production level of more than 3500 kg/ha. Such integration is suitable only for rearing and stocking ponds where fishes are above 12 gm.

Selection of duck and their maintenance:

- Some important varieties of ducks are Nageswari, Sylhet Meat, Indian Runner, Serachameli and Khaki Campbell. Out of this varieties, Khaki Campbell crossed with local pati variety is the best in Assam condition.

- It should be collected from the Government Farm and then consult with the veterinary Doctor for treatment and preventive measure of some epidemic disease like Duck Plague, Cholera etc.
- Ducks are stocked @200-300 ducklings/ha of fish pond. From duck excreta annual manure production is 45-55 kg/duck/yr, which besides fertilizing the fishponds and can be directly utilized as fish food. Apart from this, 10-20% feed/day/duck is wasted which is utilized in ponds. Duck dropping contains 81% moisture and 0.91% N and 0.38% P2O5.

Duck-house:

- Duck house should be made on the pond dyke or over the pond surface with the help of locally available material such as bamboo cane, thatches etc.
- In case of duck house over the pond surface, a small bamboo bridge is constructed from the duck house for feeding the ducks as well as for collecting eggs and duck from the house. Another bridge is constructed from the duck house to the pond surface for helping the ducks ascend or descend to pond water.
- Again duck house should be well ventilated for air circulation and exposed to direct sunlight. Periphery of the pond should be fenced for protection of ducks.
- Ducks are kept in duck house providing about 0.3-0.5 m2/bird. Again one male duck should be kept in every 5-6 female ducks. Ducklet especially up to 3-4 weeks old are very susceptible to disease, hence, care should be taken within this period.

Duck Feed:

- Mostly fine rice bran and poultry feed (layers mash etc.) are used as duck feed at the rate of 100 gm feed/day/duck.
- Duck feed should be stored at cool and dry environment. If possible, manganese sulphate mixed with feed gives the better result at the rate of 10gms/100kg of feed. Apart from that Duck weeds (Lemna, Wolfia, Azolla etc.) are also fed to the ducks.
- Duck also consume tadpoles, juvenile of frogs and dragon fly larvae. Further protein content in natural food organisms of the ponds consumed by duck is high. Therefore, the duck reared in fishponds save the cost on protein substantially in duck feeds and gives more eggs in comparison to duck which are not allowed in fishponds. The left over feed given to the ducks and duck dropping fulfil more than 59% of food requirements of farm fishes.
- Sometime algal bloom may increase, then duck dropping should be stopped periodically by using plastic at the bottom of the duck house and removed.

Lime Application

In this integration except basal manure there is no need to apply inorganic and organic manure from the outside during the culture period. Only lime is applied at the rate of 250-300 kg/ha/year at suitable intervals.

Production

By this integration a production of 3500-4000 kg of fish, 18000-18500 eggs and 500-600 kg duck meat from 1 ha of pond area in 1 year without any supplementary feed and fertilizers can be obtained and the cost is turned down to 60% lesser than normal.

Integrated Livestock-Fish Farming System in Assam

Evolved on the principles of productive recycling of farm wastes, fish-livestock farming systems are recognized as highly assured technologies for fish cultivation. In these technologies, predetermined quantum of livestock waste obtained by rearing the livestock in the pond area is applied in pond to raise the fish crop without any other additional supply of nutrients.

Here two technologies are dealt with:

a. Integrated pig fish farming.

b. Integrated cattle fish farming.

Integrated Pig Fish Farming

Integrated pig-fish farming is a highly profitable fish culture system, where pigs are reared adjacent to the fish ponds, preferably on the pond embankment from where pig urine, excreta and spilled pig feeds are introduced into the pond water. In one harvest cycle of fish (one year), 2 batches of pigs are grown, 6 months each. This is direct integration system, which is a more efficient method than the indirect integration model, wherein pigs are raised elsewhere and the pig waste is manually applied to the pond daily at a pre-determined dose.

Benefits of pig-fish farming (Direct integration):

- Fish utilizes the feed spilled by pigs and their excreta, which is very rich in nutrients for fish.
- Pig dung act as a suitable substitute to pond fertilizer and supplementary feed for some of the fishes, therefore the cost of fish production is reduced by about 60%.
- No additional land is required for piggery operations.
- Cattle fodders required for pigs and grass carps are grown on the terraced pond embankments.
- Mortality of pig is greatly reduced, as pond provides much needed water for washing the pig-sties and pigs.

- The pond mucks which gets accumulated at the pond bottom due to constant application of pig dung, can be used as an excellent fertilizer for growing vegetables, other crops and cattle fodder.
- Efficient labour utilization.

Pig Husbandry Practices

Growth of pigs depends upon many factors including breed and strain, but good management contributes to the achievement of optimum production.

Construction of pig house:

- The pig house can be constructed by using locally available materials such as bamboo and thatch, but the floor must be roughly cemented (so as to be non slippery).
- An enclosed run is provided to the pen so that the pigs get enough air, sunlight and space for dunging.
- The wall should be 1.0 m in height and preferably made of bricks. The upper part of the wall should be provided with wire netting. The height of the pig sty is 1.5 m.
- The floor of the house is slightly slanted towards a drainage canal. The canal is connected to the pond. The drainage canal is provided with a diversion canal leading to a cemented pit, where the wastes are stored in the days when the pond has algal bloom. A built-in shutter is provided in the drainage canal to regulate the flow of wastes.
- Space requirement per pig is 1.5 m^2. The roof of the house may be made using thatch or asbestos. Feeding and drinking troughs are constructed alternatively inside the pen, attached to one wall. A bath tub may also be constructed attaching to one wall of the open run.

Selection of Pigs

Pigs with 75% or 50% pure Hempshire blood has been found to be the best for such system. Landrace, large-black, etc. can also be used but not the indigenous varieties. 2-3 months old weaned piglets are brought to the pig sty for six month rearing.

Methods of preparation of pig feed using the ingredients:

- Colocasia leaves, stem and rhizomes are cut into small pieces and mixed with the above ingredients except rice bran and jubili and boiled properly. Prior to serving, jubili and rice bran are mixed.
- Feeding rate : Pigs are fed at ad-libitum. Pigs are fed twice a day i.e. morning and afternoon. In addition, pigs are to be fed regularly with wilted water hyacinth regularly. Cattle fodder, tapioca leaves, banana plants should be cut into small pieces and serve to the pigs.

Health care:

- Pig sties need be washed regularly two times a day in the summer months and once in winter season. Pigs should also be given bath twice a day in summer and once in winter.
- Disinfection of pig sties should be done twice in a week, with quick lime, and potash ($KmNO4$) respectively. The washed water leads to a fish pond serving double purpose.
- Piglets must be vaccinated against swine fever. They are to be dewormed at the age of 3-4 months.

Disposal

After rearing for about 6 months, pigs attain slaughter maturity size (60-70 kg live weight). These are to be sold out and the new piglets are to be introduced into the pigsty.

Prestocking Management

Clearance of unwanted fishes, deweeding and liming as in the case of composite fish culture system is done prior to stocking of pond with fish seed.

Stocking:

- In direct integration system, stocking is done after 20 days of introduction of piglets in the pigsty. In indirect integration, 1000 kg/ha pig wastes is applied in single installment. Pond becomes ready for stocking after 20 days.
- Rate of stocking of fish seed is 9000 nos./ha preferably with advanced fingerlings.

Species	*Percentage*	*Number/ha.*
Catla	20	1800
Rohu	20	1800
Mrigal	20	1800
Silver carp	25	2250
Grass carp	10	900
Common carp	5	450
Total	100	9000

Stock is replenished after partial harvesting with a same number of fingerlings of harvested species. After one year, complete harvesting is done.

Pond is dewatered partially to expose the silt to sunlight. These are to be taken out with the help of bamboo basket. New stock of fish is introduced after proper liming. In second year too, partial dewatering followed by desilting is to be done. At the end of third year, complete dewatering is done to expose the bottom. Complete desilting is done, followed by liming. Thereafter, fish culture is started, after 20 days of introduction of the piglets into pigsties.

Precautions :

- Monitoring of dissolved oxygen (DO) in the morning is a must particularly in the summer months. DO falling below 3.0 mg/litre particularly in pre and postmonsoon should be regarded as a warning to control further application of pig dung.
- 2. In large ponds (=0.5 ha), pig dung should not be allowed to fall on a single spot. The collected dung should be divided into 5-6 parts and applied in prespecified sites selected zones.
- If extensive algal bloom appears on water surface, pig dung, instead of introducing into pond it should be kept collected in the cemented pit.
- Stock is replenished after partial harvesting with a same number of fingerlings of harvested species. After one year, complete harvesting is done.
- Pond is dewatered partially to expose the silt to sunlight. These are to be taken out with the help of bamboo basket. New stock of fish is introduced after proper liming. In second year too, partial dewatering followed by desilting is to be done. At the end of third year, complete dewatering is done to expose the bottom. Complete desilting is done, followed by liming. Thereafter, fish culture is started, after 20 days of introduction of the piglets into pigsties.

Precautions :

- Monitoring of dissolved oxygen (DO) in the morning is a must particularly in the summer months. DO falling below 3.0 mg/litre particularly in pre and postmonsoon should be regarded as a warning to control further application of pig dung.
- 2. In large ponds (=0.5 ha), pig dung should not be allowed to fall on a single spot. The collected dung should be divided into 5-6 parts and applied in prespecified sites selected zones.
- If extensive algal bloom appears on water surface, pig dung, instead of introducing into pond it should be kept collected in the cemented pit.

Income

Production from direct integration per ha water area :-

Fish : 3000-3500 kg/ha.

Pig : 4000-5000 kg (live weight)/80 pigs/ha.

Integrated Poultry— Fish Farming System

In this system, the fish crop is integrated using only poultry droppings or dip litter by rearing the poultry either directly over the pond or on the pond embankment.

By adopting this technology, production of 3500 to 4000 kg fish, more than 20000 eggs and about 1250 kg (live weight) chicken meat can be obtained from a hectare of pond area in one year.

Poultry Husbandry Practices

In integrated poultry fish farming, the birds are confined to the house entirely, with no access to the land outside. This intensive system is of two types, viz. Battery system (Cage system) and dip litter system:

- The dip litter system is preferred over the cage system due to higher manurial value of the built up dip litter.
- In this system, the poultry birds are kept in pens up to 250 birds per pen on floor covered with litter.
- For starting the dip litter system, the floor of the pen is covered with dry organic material. The chopped straw, dry leaves, hay, saw dust etc. to a depth of about 6 inches. 0.3 to 0.4 m floor space is required per bird.
- The dropping of the birds which fall on the litter gradually combine with the litter material due to bacterial action.
- When the depth of litter becomes less, more organic matter is added to maintain sufficient depth.
- In case the litter becomes damp superphosphate or lime is added to keep it dry.
- The litter is regularly stirred for aeration. In about 10-12 months, it becomes fully built up litter, having very high manurial value.

Selection of Birds

The fowls of Rhode island, white leghorn or kuroiler are suitable for the purpose. About 500 to 600 birds (lay eggs) are required for one hectare water spread area. About eight week old chicks, after vaccination against viral diseases and providing other necessary prophylactic measures as a safeguard against epidemics are kept in poultry house near the pond.

Feeding

Grower mash is provided to the farmed birds during the age of 9-20 weeks @50-70 gm/bird/day, whereas a layer mash is provided to the birds above 20 weeks @ 80-120 gm/bird/day. The feed is provided to the birds in feed hoppers to avoid wastage. An ample supply of water is made available to all the birds at all the time.

Egg Laying

Each pen of laying birds is provided with one nest for 5-6 birds. Egg production commences at the age of 22 weeks and then gradually decline. The birds lay from 240-250 eggs per year. After the age of 18 month birds are disposed.

Health Care

The poultry house and equipment must be disinfected atleast 30 days prior to bringing in the new flock. The birds are to be vaccinated against diseases like infectious bronchitis infected laryngo trancheitis, mark's diseses, ranikhet diseses, fowl pox, etc. at the appropriate age. Some of the bacterial diseases viz. salmonellosis, coryza, fowl cholera, etc. can be kept under control by maintenance of proper hygienic conditions. Broad spectrum antibiotics may be added to the water in case the infection is mild. In severe cases, veterinary expert should be consulted.

The fully built up dip litter removed from poultry pens is stored in suitable place and supplied to the pond @50 kg/ha/day every morning after sunrise. The application of this is deferred on the days when algal bloom appears in the pond.

Production

Fresh fish : 3500 to 4000 kg/ha/yr

Egg : 70000 nos./pen

Poultry meat : 1250 kg (live weight) per year.

Rice-Fish farming in Assam

Rice fish farming can contribute to household income, contribute to food security and nutrition and contribute to improved sustainability of rice production. Generally two production systems have been recommended for culturing fish in the rice fields. They are, Simultaneous or Concurrent Method and Alternate or Rotational Method.

Selection criteria of rice field:

- The field must hold water continuously for several months. The field should be hold water to a depth of 30 cm.
- The plot should be comparatively flat
- The land should be selected keeping in view that it will not be over flooded.
- The soil of the land should hold much water. Clay soil is better for this purpose and the soil pH should be around 6.5-7.5
- Those fields where water remains even after harvest of paddy that is, there should be water, which is sufficient to continue fish culture.
- Fishes can be grown unaffected by chemicals or pesticides used to protect paddy.
- In land with slopes, a high dyke on the uphill side at the field is usually not needed. The layout of the land will help to confine the fish, if the rice field is basin shaped. This can save a lot of work because middle of the field is deepest portion and a very little effort is required to raise the dykes.
- The paddy field should have strong dykes to prevent leakage and to retain water up to desired level/depth.

- The plot should be at the close vicinity of the farmer's house so that better care can be taken up.

Selection criteria of fish species:

- The fish should be capable of tolerating very shallow water level.
- Should withstand higher temperature (up to 40°C) and variable temperature fluctuations
- Those species, which can withstand fairly high turbidity of water and poor oxygen concentration.
- The species, which have faster growth rate and should have desirable characteristics to grow to marketable size in short duration at the time of harvesting the rice.

Selection of rice varieties:

- The variety of paddy should be high yielding
- The variety should be highly disease resistant and susceptible to less attack from pests.
- Local paddy varieties of medium to long duration with non lodging characteristics are suitable.

Preparation of Fields and other Considerations

Dyke Construction

Embankments should have a height of 40-50 cm. Since water level for rice does not exceed 20cm, such embankments will already have a free board of 20-30cm. This is sufficient to prevent fish from jumping over.

Provision of Weirs and Screens

Three types of screens can be provided to prevent escape of fish and to prevent entry of predatory fishes to the plot: e.g. bamboo slats, a basket, and a piece of fish net materials even a well perforated piece of sheet metal.

Provisions of Proper Drains

The common practice is to temporarily breach a portion of embankment for water to get in or out and once the purpose is finished the breach portion be repaired. Bamboo tubes, hollowed out logs, metal pipes or bamboo chutes are also used.

Depending upon the slope of the land three types of layout for construction of rice fish culture plot has been proposed :

- Perimeter trench model,
- Central pond model,
- Lateral trench model.

Perimeter Rench Model

In this trenches are dug out in the periphery of the paddy field and the paddy cultivation area remains in the middle of the plot in zone of moderate elevation. In a plot of 1 ha, the area in the central part of the field for paddy cultivation is about 0.67 ha.

The perimeter trenches may occupy about 0.2 ha and perimeter dykes may occupy another 0.12 ha. Design and construction is that the trench is about 6 meter at the top, 3.5 m at the base and depth is 1.2m. The perimeter may be 1 m in wide at the crest and 3 m at the base.

Central Pond Model

In this case fish culture area remains at the center or at the middle of the plot and the paddy growing areas lies surrounding the pond. In a plot of 1 ha area, part of the field for paddy cultivation is about 0.65 ha and the area for the fishpond is 0.35ha with peripheral dyke space of 0.002 ha. The dimensions of perimeter dyke may be 20cm in width, base is of 50 cm width and height is of 30 cm.

Lateral Trench Model

Trapezoidal trenches are dug at the sides of paddy plot of dimensions top width 18 m, base 15m and depth is of 1.5 m. There is also to be a peripheral dyke like other two previous cases. In a plot of 1 ha area; part of the field for paddy cultivation is about 0.61 ha and the area for the fishpond is 0.27ha with peripheral dyke space of 0.12 ha.

Water Management

Continuous flooding up to the maximum tolerated by rice without affecting its rice production is recommended and it is generally 15-20 cm. At that depth, the effective water depth of 65-70 cm is available to the fish in refuge. This is sufficient to provide the fishes a cooler place when the shallow water over the rice warms up to as high as 40°C. The increased depth means a greater volume of water for rice–fish culture.

Fertilization

Application of fertilizers increases the nutrient concentration in water, which flourish the growth of phytoplankton, which are directly consumed by the fish or support good growth of zooplankton. Compost fertilizer for deep-water variety @30 tons per ha can be applied. In organic fertilizers for deep water variety Paddy as Nitrogen @120kg/ha, P205 @ 60 kg/ha and K20 @ 60 kg/ha can be applied.

Stocking Patters

Rice fish culture may involve the stocking of young fry for the production of fingerlings (nursery operation) or the growing of fingerlings to marketable size (growing operation). Rice fish farming may either be the culture of only one species (monoculture) or a combination of two or more species of fish and crustaceans (polyculture). Generally, stocking density of fish depends on its size, species and the fertility of the land. It is best to wait until the rice is well established before releasing fish seed. Fish can be stocked once two or three tillers have appeared for which the usual waiting is 1-3 weeks after transplanting or 4-6 weeks after seeding.

Feeding

To boost the growth of fish, rice bran and mustard oil cake in the ratio of 70:30 @2-3% of the body weight of fishes can be provided.

10

Oyster Management

The word oyster is used as a common name for a number of distinct groups of bivalve molluscs which live in marine or brackish habitats. The valves are highly calcified.

Some kinds of oyster are commonly consumed by humans, cooked or raw. Other kinds, such as pearl oysters, are not.

Types

True oysters

True oysters are members of the family Ostreidae. This family includes the edible oysters, which mainly belong to the genera Ostrea, Crassostrea, Ostreola and Saccostrea. Examples include the Belon oyster, Eastern oyster, Olympia oyster, Pacific oyster, Sydney rock oyster and the Wellfleet oyster.

True oysters are incapable of making gem quality pearls, although illustrations or photographs sometimes mistakenly pair them with a gem-quality pearl.

Pearl Oysters

Almost all shell-bearing molluscs can secrete pearls, yet most are not very valuable.

Pearl oysters are not closely related to true oysters, being members of a distinct family, the feathered oysters (Pteriidae). Both cultured pearls and natural pearls can be obtained from pearl oysters, though other molluscs, such as the freshwater mussels, also yield pearls of commercial value.

The largest pearl-bearing oyster is the marine Pinctada maxima, which is roughly the size of a dinner plate. Not all individual oysters produce pearls naturally. In fact, in a harvest of three tons of oysters, only three to four oysters produce perfect pearls.

In nature, pearl oysters produce natural pearls by covering a minute invading parasite with nacre, not by ingesting a grain of sand. Over the years, the irritating object is covered with enough layers of nacre to form what is known as a pearl. There are many different types, colours and shapes of pearl; these qualities depend on the natural pigment of the nacre, and the shape of the original irritant.

Pearl farmers can culture a pearl by placing a nucleus, usually a piece of polished mussel shell, inside the oyster. In three to six years, the oyster can produce a perfect pearl. These pearls are not as valuable as natural pearls, but look exactly the same. In fact, since the beginning of the 20th century, when several researchers discovered how to produce artificial pearls, the cultured pearl market has far outgrown the natural pearl market. Natural pearls have become increasingly scarce, and a necklace with only natural pearls can easily cost several hundred thousand US dollars.

Other Types of Oysters

A number of bivalve molluscs other than edible oysters and pearl oysters also have common names that include the word "oyster", usually because they either taste or look like oysters, or because they yield noticeable pearls. Examples include:

- *Thorny Oysters (*Spondylidae*)*
- Pilgrim oysters (a kind of scallop)
- *Saddle oysters (*Anomia ephippium*).*

Anatomy

Oysters are filter feeders, drawing water in over their gills through the beating of cilia. Suspended plankton and particles are trapped in the mucus of a gill, and from there are transported to the mouth, where they are eaten, digested and expelled as faeces or pseudofaeces. Oysters feed most actively at temperatures above 10 °C (50 °F). An oyster can filter up to 5 litres (1.3 US gal) of water per hour. Chesapeake Bay's once flourishing oyster population historically filtered excess nutrients from the estuary's entire water volume every three to four days. Today that would take nearly a year. Excess sediment, nutrients, and algae can result in the eutrophication of a body of water. Oyster filtration can mitigate these pollutants.

In addition to their gills, oysters can also exchange gases across their mantle, which is lined with many small, thin-walled blood vessels. A small, three-chambered heart, lying under the adductor muscle, pumps colourless blood to all parts of the body. At the same time, two kidneys, located on the underside of the muscle, remove waste products from the blood.

While oysters have two sexes, they may change sex one or more times during their life span. Because of this, it is technically possible for an oyster to fertilize its own egg. The gonads surround the digestive organs, and are made up of sex cells, branching tubules and connective tissue. Oysters cannot be sexed by examining the shell.

Habitat and Behaviour

As a keystone species, oysters provide habitat for many marine species. Crassostrea and Saccostrea live mainly in the intertidal zone, while Ostrea are subtidal. The hard surfaces of oyster shells and the nooks between the shells provide places where a host of small animals can live. Hundreds of animals such as sea anemones, barnacles, and hooked mussels inhabit oyster reefs. Many of these animals are prey to larger animals, including fish such as striped bass, black drum and croakers.

An oyster reef can increase the surface area of a flat bottom 50-fold. An oyster's mature shape often depends on the type of bottom to which it is originally attached, but it always orients itself with its outer, flared shell tilted upward. One valve is cupped and the other is flat.

Oysters usually reach maturity in one year. They are protandric; during their first year they spawn as males by releasing sperm into the water. As they grow over the next two or three years and develop greater energy reserves, they spawn as females by releasing eggs. Bay oysters usually spawn by the end of June. An increase in water temperature prompts a few oysters to spawn. This triggers spawning in the rest, clouding the water with millions of eggs and sperm. A single female oyster can produce up to 100 million eggs annually. The eggs become fertilized in the water and develop into larvae, which eventually find suitable sites, such as another oyster's shell, on which to settle. Attached oyster larvae are called spat. Spat are oysters less than 25 millimetres (0.98 in) long. Many species of bivalve, oysters included, seem to be stimulated to settle near adult conspecifics.

Some tropical oysters in the family Isognomonidae grow best on mangrove roots. Low tide can expose them, making them easy to collect. In Trinidad in the West Indies, tourists are often astounded when they are told that in the Caribbean, "oysters grow on trees."

Common oyster predators include crabs, sea birds, sea stars, and humans. Some oysters contain live crabs, known as oyster crabs.

Marine Pollution

Oysters consume nitrogen-containing compounds (nitrates and ammonia), removing them from the water. Nitrogen compounds are important phytoplankton nutrients. Phytoplankton increase water turbidity. Limiting the amount of phytoplankton in the water improves water quality and other marine life by reducing competition for dissolved oxygen. Oysters feed on plankton, incidentally consuming nitrogen compounds as well. They then expel solid waste pellets which decompose into the atmosphere as nitrogen. In Maryland, the Chesapeake Bay Program plans to use oysters to reduce the amount of nitrogen compounds entering the Chesapeake Bay by 19,000,000 pounds (8,600,000 kg) per year by 2010.

Human History

Middens testify to the prehistoric importance of oysters as food. In the United Kingdom, the town of Whitstable is noted for oyster farming from beds on the Kentish Flats that have been used since Roman times. The borough of Colchester holds an annual Oyster Feast each October, at which "Colchester Natives" (the native oyster, Ostrea edulis) are consumed.

The United Kingdom hosts several other annual oyster festivals, for example Woburn Oyster Festival is held in September. Many breweries produce Oyster Stout, a beer intended to be drunk with oysters that sometimes includes oysters in the brewing process.

The French seaside resort of Cancale is noted for its oysters, which also date from Roman times. Sergius Orata of the Roman Republic is considered the first major merchant and cultivator of oysters. Using his considerable knowledge of hydraulics, he built a sophisticated cultivation system, including channels and locks, to control the tides. He was so famous for this that the Romans used to say he could breed oysters on the roof of his house.

The world-famous Clarenbridge and Galway Oyster Festivals take place in Galway, Ireland each September. In Ireland it is traditional to eat them live with Guinness and buttered brown soda bread.

In the early 1800s, oysters were cheap and mainly eaten by the working class. Throughout the 19th century, oyster beds in New York harbor became the largest source of oysters worldwide. On any day in the late 19th century, six million oysters could be found on barges tied up along the city's waterfront. Oysters were naturally quite popular in New York City, and helped initiate the city's restaurant trade. New York's oystermen became skilled cultivators of their beds, which provided employment for hundreds of workers and nutritious food for thousands.

Eventually, rising demand exhausted many of the beds. To increase production, they introduced foreign species, which brought disease, when combined with effluent and increasing sedimentation from erosion, which destroyed most of the beds by the early 20th century. Oysters' popularity has put an ever-increasing demands on wild oyster stocks. This scarcity increased prices, converting them from their original role as working class food to their current status as an expensive delicacy.

In the United Kingdom, the native variety is still held to be the finest, requiring five years to mature and protected by an Act of Parliament during the May-August spawning season. The current market is dominated by the larger Pacific oyster and rock oyster varieties which are farmed year round.

Commercial Fishing

Fishing from the Wild

Oysters are harvested by simply gathering them from their beds. In very shallow waters they can be gathered by hand or with small rakes. In somewhat deeper water, long-handled rakes or oyster tongs are used to reach the beds. Patent tongs can be lowered on a line to reach beds that are too deep to reach directly. In all cases the task is the same: the oysterman scrapes oysters into a pile, and then scoops them up with the rake or tongs.

In some areas a scallop dredge is used. This is a toothed bar attached to a chain bag. The dredge is towed through an oyster bed by a boat, picking up the oysters in its path. While dredges collect oysters more quickly, they heavily damage the beds, and their use is highly restricted. Until 1965 Maryland limited dredging to sailboats, and even since that date motor boats can be used only on certain days of the week. These regulations prompted the development of specialized sailboats (the bugeye and later the skipjack) for dredging.

In any case, when the oysters are collected, they are sorted to eliminate dead animals, bycatch (unwanted catch), and debris. Then they are taken to market where they are either canned or sold live.

Cultivating Oysters

Oysters have been cultured for well over a century. Two methods are commonly used, release and bagging. In both cases oysters are cultivated onshore to the size of spat, when they can attach themselves to a substrate. They may be allowed to mature further to form seed oysters. In either case they are then placed in the water to mature.

The release technique involves distributing the spat throughout existing oyster beds allowing them to mature naturally to be collected like wild oysters. Bagging has the cultivator putting spat in racks or bags and keeping them above the bottom. Harvesting involves simply lifting the bags or rack to the surface and removing the mature oysters. The latter method prevents losses to some predators, but is more expensive.

The Pacific or Japanese oyster, Crassostrea gigas has been grown in the outflow of mariculture ponds. When fish or prawn are grown in ponds, it takes, typically 10 kilograms (22 lb) of feed to produce 1 kilogram (2.2 lb) of product (dry-dry basis). The other 9 kilograms (20 lb) goes into the pond and after mineralization, provides food for phytoplankton, which in turn feeds the oyster.

To prevent spawning, sterile oysters are now cultured by crossbreeding tetraploid and diploid oysters. Because the resulting triploid oyster cannot propagate, spawning does not occur preventing introduced oysters from spreading into unwanted habitats.

Restoration and Recovery

In many areas non-native oysters have been introduced in attempts to prop up failing harvests of native varieties.

For example, the eastern oyster was introduced to California waters in 1875, while the Pacific oyster was introduced there in 1929. Proposals for further such introductions remain controversial.

The Pacific oyster prospered in Pendrell Sound where the surface water is typically warm enough for spawning in the summer. Over the following years, spat spread out sporadically and populated adjacent areas. Eventually, possibly following adaptation to the local conditions, the Pacific oyster spread up and down the coast and now is the basis of the North American west coast oyster industry. Pendrell Sound is now a reserve that supplies spat for cultivation.

Near the mouth of the Great Wicomico River in the Chesapeake Bay, five year-old artificial reefs now harbor more than 180 million native Crassostrea virginica. That is still a far cry from the late 1880s, when the Bay's population was in the billions, and watermen harvested about 25 million imperial bushels (910,000 m^2) annually. The 2009 harvest was less than 200,000 imperial bushels (7,300 m^3). Researchers claim that the keys to the project were:

- using waste oyster shells to elevate the reef floor 10–18 inches (0.25–0.46 m) to keep the spat free of bottom sediments
- building larger reefs, ranging up to 20 acres (8.1 ha) in size
- disease resistant broodstock.

In 2005, China accounted for 80% of the global oyster harvest. Within Europe, France remained the industry leader.

Oysters as Food

Jonathan Swift is quoted as having said, "He was a bold man that first ate an oyster", but evidence of oyster consumption goes back into prehistory, evidenced by oyster middens found worldwide. Oysters were an important food source in all coastal areas where they could be found, and oyster fisheries were an important industry where they were plentiful. Overfishing and pressure from diseases and pollution have sharply reduced supplies, but they remain a popular treat celebrated in oyster festivals in many cities and towns.

Health Benefits of Eating Oysters

Oysters, especially 'wild', are excellent sources of several minerals, including iron, zinc and selenium, which are often low in the modern diet. They are also an excellent source of Vitamin B12. Oysters are considered the healthiest when eaten raw on the half shell.

Preparation and Storage

Oysters can be eaten on the half shell, raw, smoked, boiled, baked, fried, roasted, stewed, canned, pickled, steamed, broiled or used in a variety of drinks. Preparation widely varies. It can be as simple as opening the shell and eating the contents, including juice. Butter and salt are often added. In the case of oysters Rockefeller, preparation can be very elaborate. They are sometimes served on edible seaweed, such as brown algae.

Perhaps the definitive work on oysters as food is Consider the Oyster, by M. F. K. Fisher.

Oysters are low in food energy; one dozen raw oysters contain approximately 110 kilocalories (460 kJ), and are rich in zinc, iron, calcium, and vitamin A.

Unlike most shellfish, oysters can have a fairly long shelf-life: up to two weeks; however, their (decreasingly pleasing) taste reflects their age. Oysters should be refrigerated out of water, not frozen and in 100% humidity. Oysters stored in water under refrigeration will open, consume available oxygen and die. Care should be taken when consuming oysters. Purists insist on eating them raw, with no dressing save perhaps lemon juice, vinegar (most commonly shallot vinegar), or cocktail sauce. Upscale restaurants pair raw oysters with a home-made Mignonette sauce, which consists primarily of fresh chopped shallot, mixed peppercorn, dry white wine and lemon juice or sherry vinegar. Like fine wine, raw oysters have complex flavours that vary greatly among varieties and regions: sweet, salty, earthy, or even melon. The texture is soft and fleshy, but crisp on the palate. North American varieties include: Kumamoto and Yaquina Bay from Washington State, Malpeque from Prince Edward Island, Canada, Blue Point from Long Island, New York, and Cape May oysters from New Jersey. Salinity, mineral, and nutrient variations in the water that nurtures them influence their flavor profile.

Non-local oysters are generally expensive. In the United States, oysters are most often cooked, but there is also a high demand for raw oysters on the half-shell (shooters) at oyster bars. Canned smoked oysters are also widely available as preserves with a long shelf life. Raw oysters are still found in many temperate areas bordering a sea or ocean. Oysters are commonly eaten raw in France in bars and as a 'bar fast food' but the home use tends to be mixed with a large usage in cooking-steamed or in paella or soups.

It was once assumed that oysters were only safe to eat in months with the letter 'r' in their English and French names. This is a myth whose basis in truth is that in the northern hemisphere oysters are much more likely to spoil in May, June, July, and August.

Oysters must be eaten alive, or cooked alive. The shells of live oysters are normally tightly closed or snap shut given a slight tap. If the shell is open, the oyster is dead, and cannot be eaten safely.

Cooking oysters in the shell kills the oysters and causes them to open by themselves. Oysters that don't open were dead before cooking and are unsafe.

Oysters can contain harmful bacteria. Oysters are filter feeders and will naturally concentrate anything present in the surrounding water. Oysters from the Gulf Coast of the United States, for example, contain high bacterial loads of human pathogens in the warm months, most notably Vibrio vulnificus and Vibrio parahaemolyticus. In these cases, the main danger is for immuno-compromised individuals, who are unable to fight off infection and can succumb to septicemia, leading to death. Vibrio vulnificus is the most deadly seafood-borne pathogen, with a higher case-to-death ratio than even Salmonella enterica and Escherichia coli.

Oysters are sometimes cited as an aphrodisiac. It is disputed whether this is true. A team of American and Italian researchers analyzed bivalves and found they were rich in rare amino acids that trigger increased levels of sex hormones. Also oysters have a high zinc content, a mineral that aids in the production of testosterone. Alternatively, the oyster's erotic reputation may only be due to its soft, moist texture and appearance.

Opening Oysters

Fresh oysters must be alive just before consumption or cooking. There is a simple criterion: the oyster must be capable of tightly closing its shell. Open oysters should be tapped on the shell: a live oyster will close up and is safe to eat. Oysters which are open and unresponsive are dead and must be discarded. Some dead oysters, or oyster shells which are full of sand may be closed. These make a distinctive noise when tapped, and are known as clackers.

Opening oysters requires skill. The preferred method is to use a special knife (called an oyster knife, a variant of a shucking knife), with a short and thick blade about 5 centimetres (2.0 in) long.

Insert the blade, with moderate force and vibration if necessary, at the hinge between the two valves. Then twist the blade until there is a slight pop. Then slide the blade upward to cut the adductor muscle which holds the shell closed. Inexperienced shuckers can apply too much force, which can result in injury if the blade slips. Heavy gloves are necessary: apart from the knife, the shell itself can be razor sharp. Professional shuckers require less than 3 seconds to do the deed.

If the oyster has a particularly soft shell, the knife can be inserted instead in the sidedoor, about halfway along one side where the oyster lips widen and there is a slight indentation.

Opening, or "shucking" oysters has become a competitive sport. Oyster shucking competitions are staged around the world. Widely acknowledged to be the premiere event, the Guiness World Oyster Opening Championship is held in September at the Galway Oyster Festival, in Galway Ireland.

Ethical Considerations

While technically an animal, the oyster is considered by some ethicists to be an appropriate food choice for vegans and vegetarians, arguing it is acceptable to eat oysters, because in the relevant ethical terms they are rather closer to plants than animals. Two common ethical objections to the consumption of animals is that they feel pain (and that causing pain is wrong), and that their cultivation is environmentally harmful. On both of these, oysters are significantly closer to plants than animals. Regarding pain, oysters lack a central nervous system, and are not believed to experience pain in the same way as humans do, with they and other bivalves being closer to mobile plants than to plant perception. Regarding environmental impact, 95% of oysters are sustainably farmed and harvested (other bivalves are frequently harvested by harmful dredging), feed on plankton (very low on the food chain), and in fact improve the marine environment by removing toxins. As such, oysters are listed as a "Best Choice" (highest rating) on the Seafood Watch list. The view that oysters are acceptable to eat, even by strict ethical criteria, has notably been propounded in the seminal 1975 text Animal Liberation, by philosopher Peter Singer; while subsequent editions have reversed this position (advocating against eating oysters), Singer has stated that he has "gone back and forth on this over the years," and as of 2010 states that he sees "no good reason for avoiding eating sustainably produced oysters".

Diseases

Oysters are subject to various diseases which can reduce harvests and severely deplete local populations. Disease control focuses on containing infections and breeding resistant strains and is the subject of much ongoing research.

- Dermo (Perkinsus marinus) is caused by a protozoan parasite. It is a prevalent pathogen, causes massive mortality and poses a significant economic threat to the oyster industry. The disease is not a direct threat to humans consuming infected oysters. Dermo first appeared in the Gulf of Mexico in the 1950s, and until 1978 it was believed to be caused by a fungus. While it is most serious in warmer waters, it has gradually spread up the east coast of the United States.
- MSX (Multinucleated Sphere X) is caused by the protozoan Haplosporidium nelsoni, generally seen as a multi-nucleated plasmodium. It is infectious and causes heavy mortality in the Eastern Oyster; survivors, however, develop resistance and can help propagate resistant populations. MSX is associated with high salinity and water temperatures. MSX was first noted in Delaware Bay in 1957 and is now found all up and down the east coast of the United States. Evidence suggests that it was brought to the United States when Crassostrea gigas, a Japanese oyster variety, was introduced to Delaware Bay.

Freshwater Pearl Culture

What is Pearl Culture?

Pearl is a natural gem and is produced by a mollusc. While the demand of pearls in India and elsewhere is increasing, their supplies from nature have reduced due to over exploitation and pollution. India is importing a large amount of cultured pearls every year from international market to meet the domestic demand. The Central Institute of Freshwater Aquaculture, Bhubaneswar has developed the technology of Freshwater Pearl Culture from common freshwater mussels, which are widely distributed in freshwater habitats throughout the country.

In nature, a pearl is formed when a foreign particle viz., piece of sand, insects, etc. by chance enters into the body of mussel and the mussel can not reject that out and instead makes a shiny coating on the particle layer by layer. This simple phenomenon is being exploited in pearl culture practices.

The pearl is similar to the inner shining layer of shell called 'mother of pearl layer' or nacre, constituted by calcium carbonate, organic matrix and water. The pearls available in the market could be artificial, natural or cultured. Artificial or imitation pearls are not pearls but pearl-like materials that simply contain a rigid, round core or base and an outer pearly coating. In natural pearls the core or nucleus is minute with thick pearl nacre. Generally, a natural pearl is small in size and irregular in shape. A cultured pearl is also a natural pearl, the only difference being the human intervention in surgical implantation of a live mantle graft and nucleus for hastening pearl formation to the desired size, shape, colour and lusture. In India, three species of commonly available freshwater mussels viz., Lamellidens marginalis, L. corrianus and Parreysia corrugata found to produce good quality pearls.

Culture Practices

Farming practice of the freshwater pearl culture operation involves six major steps sequentially viz., collection of mussels, pre-operative conditioning, surgery, post-operative care, pond culture and harvesting of pearls.

Collection of Mussels

The healthy mussels are collected from the freshwater bodies like pond, river etc. They are collected manually and kept in buckets or containers with water. The ideal mussel size used for pearl culture is over 8 cm in anterior-posterior length.

Pre-operative Conditioning

The collected mussels are kept for pre-operative conditioning for 2 to 3 days by keeping them in crowded condition in captivity with aged tap water at a stocking density of 1 mussel/l Pre-operative conditioning helps in weakening of adductor muscles, which helps in easy handling during surgery.

Mussel Surgery

Depending on the place of surgery the implantation is of three types viz., mantle cavity, mantle tissue and gonadal implantations. The key materials required during the surgical implantations are beads or nuclei, which are usually made from mollusc shell or other calcareous materials.

Mantle cavity implantation: In this procedure round (4-6 mm diameter) or designed (images of Ganesh, Buddha, etc.) beads are inserted into the mantle cavity region of mussel after opening the two valves (without causing injury to mussels at both ends) of animal and separating carefully the mantles of anterior sides from the shell with the help of surgical set. Implantation could be done in mantle cavities of both the valves. In case of implantation of designed beads care is taken such a way that the design portion faces the mantle. After placing the beads in desired place the gaps created during implantation are closed just by pushing the mantle onto the shell.

Mantle tissue implantation: Here the mussels are divided into two groups; the donor and the recipient mussels. The first step in this procedure is preparation of graft (small pieces of mantle tissue). This is done by preparing a mantle ribbon (a strip of mantle along the ventral side of the mussel) from a donor mussel, which is sacrificed, and cutting that into small pieces (2 x 2 mm). The implantation is done on recipient mussels, which is of two types viz., non-nucleated and nucleated. In the former, only the graft pieces are introduced into the pockets created at the inner side of posterior pallial mantle present at the ventral region of the mussel. In the nucleated method, a graft piece followed by a small nucleus (2 mm dia) is introduced in the pockets. In both the procedures care is taken so that graft or nucleus does not come out of the pocket. Implantations could be done at mantle ribbons of both valves.

Gonadal implantation: This procedure also involves preparation of grafts as described earlier (mantle tissue method). First a cut is made at the edge of the gonad of the mussel. Then a graft is inserted into the gonad followed by nucleus (2-4 mm dia) so that the nucleus and graft should be in close contact. Care is taken such a way that nucleus touches the outer epithelial layer of the graft and the intestine is not cut during the surgery.

Post-operative Care

Implanted mussels are kept in post-operative care unit in nylon bags for 10 days with antibiotic treatment and supply of natural food. The units are examined daily with removal of dead mussels and the ones that reject the nucleus.

Pond Culture

After post-operative care the implanted mussels are stocked in the ponds. The mussels are kept in nylon bags (2 mussels per bag) and are hung from bamboo or

PVC pipes and placed in ponds at 1 m depth. The mussels are cultured at stocking density of 20,000-30,000/ha. The ponds are fertilised with organic and inorganic fertiliser periodically to sustain the plankton productivity. Periodical checking of mussels with removal of dead ones and cleaning of bags is required throughout the culture period of 12-18 months.

Pearl Harvest

At the end of the culture period the mussels are harvested. While individual pearls can be taken out from the mantle tissue or gonad of the live mussels, the mussels are sacrificed in case of mantle cavity method. The products obtained through different surgical implantation methods are shell attached half round and shell attached image pearls in mantle cavity method; unattached small irregular or round pearls in mantle tissue method; and unattached big irregular or round pearls in gonadal method.

11

Policies and Schemes of Farms

Livestock Insurance Scheme

The Livestock Insurance Scheme, a centrally sponsored scheme, which was implemented on a pilot basis during 2005-06 and 2006-07 of the 10th Five Year Plan and 2007-08 of the 11th Five Year Plan in 100 selected districts. The scheme is being implemented on a regular basis in 300 districts of the country. The Livestock Insurance Scheme has been formulated with the twin objective of providing protection mechanism to the farmers and cattle rearers against any eventual loss of their animals due to death and to demonstrate the benefit of the insurance of livestock to the people and popularize it with the ultimate goal of attaining qualitative improvement in livestock and their products.

Under the scheme, the indegenous/crossbred milch cattle and buffaloes are being insured at maximum of their current market price. The premium of the insurance is subsidized to the tune of 50%. The entire cost of the subsidy is being borne by the Central Government. The benefit of subsidy is being provided to a maximum of 2 animals per beneficiary for a policy of maximum of three years.

The scheme is being implemented in all states except Goa through the State Livestock Development Boards of respective states.

Animals to be covered under the scheme and selection of beneficiaries:

- The indigenous/crossbred milch cattle and buffaloes will be under the purview of the scheme. Milch cattle/buffalo will include both in-milk and dry as well as pregnant animal, which have already calved once.
- Animals covered under any other insurance scheme/plan scheme will not be covered under this scheme.
- Benefit of subsidy is to be restricted to two animals per beneficiary and is to be given for one time insurance of an animal up to a maximum period of three years.

- The farmers will have to be encouraged to go for a three-year policy which is likely to be more economical and useful for getting the real benefit of insurance on occurrence of natural calamities like flood and drought etc. However, if a farmer desires to have a policy for a period less than three years that could also be provided and subsidy on premium will be provided for insuring same animals again in the future years of implementation of the scheme.

Determination of Market Price of the Animal

An animal will be insured for the maximum of its current market price. The market price of the animal to be insured will be assessed jointly by the beneficiary, authorized veterinary practitioner and the insurance agent.

Identification of Insured Animal

The animal insured will have to be properly and uniquely identified at the time of insurance claim. The ear tagging should, therefore, be fool proof as far as possible. The traditional method of ear tagging or the recent technology of fixing microchips could be used at the time of taking the policy.

The cost of fixing the identification mark will be borne by the Insurance companies and responsibility of its maintenance will lie on the concerned beneficiaries. The nature and quality of tagging materials will be mutually agreed by the beneficiaries and the Insurance Company.

Change of Owner During the Validity Period of Insurance

In case of sale of the animal or otherwise transfer of animal from one owner to other, before expiry of the Insurance Policy, the authority of beneficiary for the remaining period of policy will have to be transferred to the new owner. The modalities for transfer of livestock policy and fees and sale deed etc required for transfer, should be decided while entering into contract with the insurance company.

Settlement of Claims

In case of claim becoming due, the payment of insured amount should be made within 15 days positively after submission of requisite documents. Only four documents would be required by insurance companies for settling the claims viz.

FIR with the Insurance Company, Insurance Policy, Claim Form and Postmortem Report. While insuring the animal, CEOs must ensure that clear cut procedures are put in place for settlement of claims and the required documents are listed and the same is made available to concerned beneficiaries along with the policy documents.

All documents/forms for insuring as well as settling the claims should be made available by the insurance agency in local language and English language.

Fodder and Feed Development Scheme

The Department Animal Husbandry, Dairying and Fisheries is implementing a 'Centrally Sponsored Fodder Development Scheme' to supplement the efforts of states in feed and fodder development. The scheme is being implemented from 2005-06 with the following four components:

a. Establishment of Fodder Block Making Units
b. Grassland Development including Grass Reserves
c. Fodder Seed Production and Distribution.
d. Biotechnology Research Projects.

The Centrally Sponsored Fodder Development Scheme has been modified from 2010 to ensure efficient use of available fodder/feed. With an outlay of Rs. 141.40 crore, this scheme incorporates the following new components/technology interventions:

- Strengthening of Feed Testing Laboratories
- Introduction of Chaff cutters
- Establishment of silage making units
- Demonstration of Azolla cultivation and production units
- Establishment of by-pass protein production units
- Establishment of Area Specific Mineral Mixture (ASMM)/Feed Pelleting/ Feed manufacturing Units.

Further the amount of subsidy given under the ongoing component, Establishment of Fodder Block Making Units has been increased to 50% to encourage better participation and the land requirement for assistance under the Grassland Development including Grass Reserves component has been revised to 5-10 Ha.

Diseases of Animals

Disease is the general term for any deviation from the normal or healthy condition of the body. The morbid processes that result in either slight or marked modifications of the normal condition are recognized by the injurious changes in the structure or function of the organ, or group of body organs involved. The increase in the secretion of urine noticeable in horses in the late fall and winter is caused by the cool weather and the decrease in the perspiration. If, however, the increase in the quantity of urine secreted occurs independently of any normal cause and is accompanied by an unthrifty and weakened condition of the animal, it would then characterize disease. Tissues may undergo changes in order to adapt themselves to different environments, or as a means of protecting themselves against injuries.

The coat of a horse becomes heavy and appears rough if the animal is exposed to severe cold. A rough, staring coat is very common in horses affected by disease.

The outer layer of the skin becomes thickened when subject to pressure or friction from the harness.

This change in structure is purely protective and normal. In disease the deviation from normal must be more permanent in character than it is in the examples mentioned above, and in some way prove injurious to the body functions.

Classification

We may divide diseases into three classes: non-specific, specific and parasitic.

Non-specific diseases have no constant cause. A variety of causes may produce the same disease. For example, acute indigestion may be caused by a change of diet, watering the animal after feeding grain, by exhaustion and intestinal worms. Usually, but one of the animals in the stable or herd is affected. If several are affected, it is because all have been subject to the same condition, and not because the disease has spread from one animal to another.

Specific Diseases

The terms infectious and contagious are used in speaking of specific diseases. Much confusion exists in the popular use of these terms. A contagious disease is one that may be transmitted by personal contact, as, for example, influenza, glanders and hog-cholera. As these diseases may be produced by indirect contact with the diseased animal as well as by direct, they are also infectious. There are a few germ diseases that are not spread by the healthy animals coming in direct contact with the diseased animal, as, for example, black leg and southern cattle fever. These are purely infectious diseases. Infection is a more comprehensive term than contagion, as it may be used in alluding to all germ diseases, while the use of the term contagion is rightly limited to such diseases as are produced principally through individual contact.

Parasitic diseases are very common among domestic animals. This class of disease is caused by insects and worms, as for example, lice, mites, ticks, flies, and round and flat worms that live at the expense of their hosts. They may invade any of the organs of the body, but most commonly inhabit the digestive tract and skin. Some of the parasitic insects, mosquitoes, flies and ticks, act as secondary hosts for certain animal microorganisms that they transmit to healthy individuals through the punctures or the bites that they are capable of producing in the skin.

Causes

For convenience we may divide the causes of disease into the predisposing or indirect, and the exciting or direct.

The predisposing causes are such factors as tend to render the body more susceptible to disease or favour the presence of the exciting cause. For example, an animal that is narrow chested and lacking in the development of the vital organs lodged in the thoracic cavity, when exposed to the same condition as the other members of the herd, may contract disease while the animals having better conformation do not. Hogs confined in well-drained yards and pastures that are free from filth, and fed in pens and on feeding floors that are clean, do not become hosts for large numbers of parasites. Hogs confined in filthy pens are frequently so badly infested with lice and intestinal worms that their health and thriftiness are seriously interfered with. In the first case mentioned the predisposition to disease is in the individual, and in the second case it is in the surroundings.

Diagnosis and Symptoms of Disease

The importance of recognizing or diagnosing the seat and nature of the morbid change occurring in an organ or group of organs cannot be overestimated. Laymen do not comprehend the difficulty or importance of correctly grouping the signs or symptoms of disease in such a way as to enable them to recognize the nature of the disease. In order to be able to understand the meaning of the many symptoms or signs of disease, we must possess knowledge of the structure and physiological functions of the different organs of the body. We must be familiar with the animal when it is in good health in order to be able to recognize any deviation from the normal due to disease, and we must learn from personal observation the different symptoms that characterize the different diseases. Stockmen should be able to tell when any of the animals in their care are sick as soon as the first symptom of disease manifests itself, by changes in the general appearance and behaviour. But in order to ascertain the exact condition a general and systematic examination is necessary.

The examiner, whether he be a layman or a veterinarian, must observe the animal carefully, noting the behaviour, appearance, surroundings, and general and local symptoms. Before making a general examination of the animal it is well, if the examiner is not already acquainted with the history of the case (care, feed and surroundings), to learn as much about this from the attendant as is possible. Inquiry should be made as to the feeding, the conditions under which the animal has been kept, the length of time it has been sick, its actions, or any other information that may be of assistance in forming the diagnosis and outlining the treatment.

The general symptoms inform us regarding the condition of the different groups of body organs. A careful study of this group of symptoms enables us correctly to diagnose disease and inform ourselves as to the progress of long, severe affections. These symptoms occur in connection with the pulse, respirations, body temperature, skin and coat, visible mucous membranes, secretions and excretions, and behaviour of the animal.

The local symptoms are confined to a definite part or organ. Swelling, pain, tenderness and loss of function are common local symptoms. A direct symptom may also be considered under this head because of its direct relation to the seat of disease. It aids greatly in forming the diagnosis.

Other terms used in describing symptoms of disease are objective , which includes all that can be recognized by the person making the examination; indirect, which are observed at a distance from the seat of the disease; and premonitory , which precede the direct, or characteristic symptoms. The subjective symptoms include such as are felt and described by the patient. These symptoms are available from the human patient only.

Pulse.—The character of the intermittent expansion of the arteries, called the pulse, informs us as to the condition of the heart and blood-vessels. The frequency of the pulse beat varies in the different species of animals. The smaller the animal the more frequent the pulse. In young animals the number of beats per minute is greater than in adults. Excitement or fear, especially if the animal possesses a nervous temperament, increases the frequency of the pulse. During, and for a short time after, feeding and exercise, the pulse rate is higher than when the animal is standing at rest. The following table gives the normal rate of the pulse beats per minute:

- Horse 36 to 40 per minute
- Ox 45 to 50 per minute
- Sheep 70 to 80 per minute
- Pig 70 to 80 per minute
- Dog 90 to 100 per minute.

In sickness the pulse is instantly responsive. It is of the greatest aid in diagnosing and in noting the progress of the disease.

The following varieties of pulse may be mentioned:

- frequent,
- infrequent,
- quick,
- slow,
- large,
- small,
- hard,
- soft and
- intermittent.

The terms frequent and infrequent refer to the number of pulse beats in a given time; quick and slow to the length of time required for the pulse wave to pass beneath the finger; large and small to the volume of the wave; hard and soft to its compressibility; and intermittent to the occasional missing of a beat. A pulse beat that is small and quick, or large and soft, is frequently met with in diseases of a serious character.

Illustration: The X on the lower border of the jaw indicates the place where the pulse is taken.

The horse's pulse is taken from the submaxillary artery at a point anterior to, or below the angle of the jaw and along its inferior border. It is here that the artery winds around the inferior border of the jaw in an upward direction, and, because of its location immediately beneath the skin, it can be readily located by pressing lightly over the region with the fingers.

Cattle's pulse is taken from the same artery as in the horse. The artery is most superficial a little above the border of the jaw. It is more difficult to find the pulse wave in cattle than it is in horses, because of the larger amount of connective tissue just beneath the skin and the heavier muscles of the jaw. A very satisfactory pulse may be found in the small arteries located along the inferior part of the lateral region of the tail and near its base.

The sheep's pulse may be taken directly from the femoral artery by placing the fingers over the inner region of the thigh. By pressing with the hand over the region of the heart we may determine its condition.

The hog's pulse can easily be taken from the femoral artery on the internal region of the thigh. The artery crosses this region obliquely and is quite superficial toward its anterior and lower portion.

The dog's pulse is usually taken from the brachial artery. The pulse wave can be readily felt by resting the fingers over the inner region of the arm and just above the elbow. The character of the heart beats in dogs may be determined by resting the hand on the chest wall.

Respiration

The frequency of the respirations varies with the species. The following table gives the frequency of the respirations in domestic animals:

- Horse 8 to 10 per minute
- Ox 12 to 15 per minute
- Sheep 12 to 20 per minute
- Dog 15 to 20 per minute
- Pig 10 to 15 per minute.

The ratio of the heart beats to the respirations is about 1:4 or 1:5. This ratio is not constant in ruminants. Rumination, muscular exertion and excitement increase the frequency and cause the respirations to become irregular. In disease the ratio between the heart beats and respirations is greatly disturbed, and the character of the respiratory sounds and movements may be greatly changed.

Illustration: Auscultation of the lungs can be practised to an advantage over the outlined portion of the chest wall, only.

Severe exercise and diseased conditions of the lungs cause the animal to breathe rapidly and bring into use all of the respiratory muscles. Such forced or laboured breathing is a common symptom in serious lung diseases, "bloat" in cattle, or any condition that may cause dyspnoea. Horses affected with "heaves" show a double contraction of the muscles in the region of the flank during expiration. In spasm of the diaphragm or "thumps" the expiration appears to be a short, jerking movement of the flank. In the abdominal form of respiration the movements of the walls of the chest are limited. This occurs in pleurisy. In the thoracic form of respiration the abdominal wall is held rigid and the movement of the chest walls make up for the deficiency. This latter condition occurs in peritonitis. A cough is caused by irritation of the membrane lining the air passages.

The character of the cough may vary according to the nature of the disease. We may speak of a moist cough when the secretions in the air passages are more or less abundant. A dry cough occurs when the lining membrane of the air passages is dry and inflamed. This may occur in the early stage of the inflammation, or as a result of irritation from dust or irritating gases. Chronic cough occurs when the disease is of long duration or chronic. In pleurisy the cough may be short and painful, and in broken wind, deep and suppressed. In parasitic diseases of the air passages and lungs, the paroxysm of coughing may be severe and "husky" in character.

The odor of the expired air, the character of the discharge and the respiratory sounds found on making a careful examination are important aids in arriving at a correct diagnosis, and in studying the progress of the disease.

Illustration: Fever thermometer.

Body Temperature : The body temperature of an animal is taken by inserting the fever thermometer into the rectum. In large animals a five-inch, and in small animals a four-inch fever thermometer is used. It should be inserted full length and left in position from one and one-half to three minutes, depending on the rapidity with which it registers.

The average normal body temperatures of domestic animals are as follows:

- Horses 100.5\260 F.
- Cattle 101.4\260 F.
- Sheep 104.0\260 F.
- Swine 103.0\260 F.
- Dog 101.4\260 F.

There is a wide variation in the body temperatures of domestic animals. This is especially true of cattle, sheep and hogs. In order to determine the normal temperature of an animal, it may be necessary to take two or more readings at different times, and compare them with the body temperatures of other animals in the herd that are known to be healthy.

Exercise, feeding, rumination, excitement, warm, close stables, exposure to cold and drinking ice cold water are common causes of variations in the body temperatures of domestic animals.

Visible Mucous Membranes : The visible mucous membranes, as they are termed, are the lining membranes of the eyelids, nostrils and nasal cavities, and mouth. In health they are usually a pale red, excepting when the animal is exercised or excited, when they appear a brighter red and somewhat vascular. In disease the following changes in colour and appearance may be noted: When inflamed, as in cold in the head, a deep red; in impoverished or bloodless conditions of the body and in internal haemorrhage, pale; in diseases of the liver, sometimes yellowish, or dark red; in diseases of the digestive tract (buccal mucous membrane), coated; if inflamed, dry at first, later excessively moist; and in certain germ diseases a mottled red, or showing nodules, ulcers and scars.

Surface of the Body : When a horse is in a good condition and well caredfor, the coat is short, fine, glossy and smooth and the skin pliable and elastic. Healthy cattle have a smooth, glossy coat and the skin feels mellow and elastic. The fleece of sheep should appear smooth and have plenty of yolk, the skin pliable and light pink in colour. When the coatloses its lustre and gloss and the skin becomes hard, rigid, thickened and dirty, it indicates a lack of nutrition and an unhealthy condition of the body. In sheep, during sickness, the wool may become dry and brittle and the skin pale and rigid. When affected with external parasites, the hair or wool becomes dirty and rough, a part of the skin may be denuded of hair, and it appears thickened, leathery and scabby, or shows pimples, vesicles and sores.

During fever, the temperature of the surface of the body is very unequal. In serious diseases or diseases that are about to terminate fatally, the skin feels cold and the hair is wet with sweat. When animals are allowed to “rough it” during the cold weather, the coat of hair becomes heavy and rough. This is a provision of nature and enables them, as long as the coat is dry, to withstand severe cold.

Horses that are in a low physical condition, or when accustomed to hardwork, if then kept in a stall for a few days without exercise, commonly show a filling of the cannon regions of the posterior extremities. This condition also commonly occurs in disease and in mares that have reached the latter period of pregnancy. Sheep that are unthrifty and in a poor physical condition, especially if this is due to internal parasites, frequently develop dropsical swellings in the region of the jaw, or neck.

Body Excretions : The character of the body excretions, faeces and urinemay become greatly changed in certain diseases. It is important that the stockman or veterinarian observe these changes, and in certain diseases make an analysis of the urine. This may be necessary in order properly to diagnose the case.

Behaviour of the Animal : When the body temperature is high, the animal may appear greatly depressed. If suffering severe pain, it may be restless. In diseases of the nervous system, the behaviour of the animal may be greatly changed. Spasms, convulsions, general local paralysis, stupid condition and unconsciousness may occur as symptoms of this class of disease.

Treatment

Preventive Treatment: The subject of preventive medicine becomes more important as our knowledge of the cause of disease advances. A knowledge of feeds, methods of feeding, care, sanitation and the use of such biological products as bacterins, vaccines and protective serums is of the greatest importance to the farmer and veterinarian. We are beginning to realize that one of the most important secrets of profitable and successful stock raising is the prevention of disease; that the agricultural colleges are doing a great work in helping to teach farmers that there are right and wrong methods of feeding and caring for animals; that the practice of sanitation in caring for animals is the cheapest method of treating disease; and that it is advisable to practise radical methods of control, when necessary, in order to rid the herd of an infectious disease.

The ration fed and the method of feeding are not only important in considering the causes of diseases of the digestive tract, but diseases of other organs as well. The feeding of an excessive, or insufficient quantity of feed, or a ration that is too concentrated, bulky and in nutritious, poor in quality, or spoiled may produce disease.

An impure water supply is a common cause of disease. A deep well that is closed in properly and does not permit of contamination from filth, does not insure a clean water supply if the trough or tank is not kept clean.

Farm Buildings: If stockmen would make a more careful study of the kind of farm buildings most suitable to their needs, the selection of the location, the proportions, the arrangement of the interior and the lighting and ventilation, there would be a great saving in losses from disease, and the cost of building in many cases would be lessened. Your neighbour's building that you have taken for your model may not be suitable for your needs. It may be more expensive than your financial condition permits. It may be poorly lighted and ventilated and not suited to the site that you have selected.

Biological Products: There are a number of biological products that maybe used in the prevention and control of disease. Some of these products, such as tuberculin and male in, enable the owner to rid his herds of tubercular cows and glandered horses before these diseases have become far enough advanced to be recognized by the visible symptoms alone. Black leg, anthrax and hog-cholera vaccines are valuable agents in the control of disease. In the treatment of fistula and infectious abortion, bacterins maybe used. There are many other germ diseases and infections for which vaccines and bacterins may be used. However, we must not depend wholly on these agents in the control of disease. We must possess a knowledge of the manner in which the infection is spread, for without this knowledge we would be unable to prevent its dissemination over a wide area.

Medicinal Treatment : The average stockman or veterinarian is more familiar with the treatment of disease with drugs than he is with the preventive measures just described. This statement does not imply that a knowledge of medicinal therapeutics is not of the greatest importance in the treatment of disease. The ultimate object of all drugs is both to prevent and cure disease, but the injudicious use of a drug does neither. A discussion of this subject cannot be entered into here, and because of its largeness it is not advisable to discuss it further than a brief summary of the methods of administering drugs.

Administration of Drugs : Drugs may be administered by the following channels: by way of the mouth, in the feed or as a drench; by injecting into the tissues beneath the skin or hypodermically; by rubbing into the skin; by the air passages and the lungs; and by injecting into the rectum.

If the animal is not too sick to eat and the drug does not possess an unpleasant taste, it may be given with the feed. If soluble, it may be given with the drinking water, or in any case, it may be mixed with ground feed if this method is to be preferred. In all cases the medicine must be well mixed with the feed. This is especially important if there are a number of animals to be treated, as there is more certainty of each animal getting the proper dose and the danger of overdosing is avoided. If the young animal is nursing the mother, we can take advantage of certain drugs being eliminated in the mother's milk and administer the drug to the mother.

Drenches

In the larger animals a bulky drench is sometimes difficult to administer, and we should, in all cases, count on a portion being wasted.

Horses are sometimes difficult to drench, and it may be advisable to confine the horse in some way. Small drenches can readily be given with asyringe or a small bottle. In giving bulky drenches it is most convenient to use a long-necked, heavy glass bottle. The horse should be backed into a narrow stall and the head elevated by placing a loop in the end of a small rope over the upper jaw, passing the rope back of the nose piece on the halter and throwing it over a beam, and raising the head until the mouth is slightly higher than the throat. If the horse refuses to swallow, a tablespoonful of clean water may be dropped into the nostril. This forces it to swallow. A drench should never be given through the nose, as it may pass into the air passages and cause a fatal inflammation of the lungs.

Cattle can be easily drenched by taking hold of the nostrils with the fingers, or snapping a bull ring into the partition between the no strils and elevating the head.

Sheep may be drenched either in the standing position, or when thrown on the haunches and held between the knees. Care should be exercised in giving irritating drenches to sheep, especially if the drench be bulky.

A herd of hogs may be quickly and easily drenched if they are confined in a small pen, and the loop of a small rope placed around the snout, well back toward the corners of the mouth. A small metal dose syringe should be used. If the drench is bulky and the hog difficult to hold, it may be necessary to elevate the head and raise the forefeet from the ground. The drench should not be given until the hog is quiet and well under control, as there is some danger of the medicine passing into the air passages and doing harm. It may be necessary to mark the hogs that have been drenched with a daub of paint, or in some other manner in order to be able to distinguish them from the untreated animals.

The administration of drugs enclosed in a gelatin capsule, or mixing them with syrup, honey or linseed oil, and rolling the mass into the form of acylinder is commonly practised. The capsule or ball may then be shot into the pharynx with a balling gun. A ball may also be given to the larger animals by carrying it into the back part of the mouth with the hand, and placing it on the back part of the tongue. In the horse this method of administration requires some practice. The tongue must be pulled well forward, the head held up, and the tongue released as soon as the ball is placed on the tongue, so that it may pass back into the pharynx.

The administration of drugs by injecting beneath the skin is suitable when the drug is non-irritating and the dose is small. Drugs administered in this way act promptly and energetically. The alkaloid or active principle of the drug is commonly used. A fold of the skin is picked up with the fingers and the needle is quickly introduced, care being taken not to prick or scratch the muscular tissue, as this causes some pain and makes the animal restless. In order to avoid abscess formation at the point of injection, the skin should be cleansed with a disinfectant and the syringe and needle sterilized before using.

Drugs are not absorbed through the unbroken skin, but when applied with friction, or when the outer layer is removed by blistering, absorption may take place. Liniments, blisters and poultices are the preparations used.

Volatile drugs, such as chloroform and ether, are absorbed quickly by the enormous vascular surface of the lungs. This class of drugs is administered for the purpose of producing general anaesthesia. Anaesthetics are indispensable in many surgical operations.

The administration of a drug in the form of medicated steam is quite useful in combating some respiratory diseases. In steaming large animals apail about half full of boiling-hot water to which has been added about announce of coal-tar disinfectant, or whatever drug is required, is held within about one foot of the animal's nostrils. It is usually advisable to throw a light cover over the head and pail in order to direct the steam toward the nostrils. Dogs can be placed on a cane-seated chair and a pailor pan of boiling-hot water placed under it, and a sheet thrown over all.

Drugs are administered by way of the rectum when the animal can not bed renched, or the drug can not be given in any other way and when a localaction is desired. An enema or clyster is a fluid injection into the rectum and is employed for the following purposes: to accelerate the action of a purgative; to stimulate the peristaltic movement of the intestines; to kill intestinal parasites; to reduce body temperature; to administer medicine; and to supply the animal with food.

An enema may be administered by allowing water to gravitate into the rectum from a height of two or three feet or by using an injection pump. In the larger animals several feet of heavy walled rubber tubing carrying a straight nozzle at one end should be used. In administering an enema, the rectum should be emptied outwith the hand and the nozzle of the syringe carried as far forward as possible. The operator should be careful not to irritate or tear the wall of the rectum.

Size of the Dose : The doses recommended in the treatment of the different diseases, unless otherwise stated, are for mature animals. The dose for a colt one year of age is about one-third the quantity given the adult, two years of age one-half, and three years of age two-thirds. In well-matured colts a larger dose may be given. In cattle, the doses recommended are about the same. In the smaller animals the size of the dose may be based on the development and age of the animal. When the drug is administered at short intervals or repeated, the size of the dose should be reduced. The physiological action of some drugs may be changed by varying the size of the dose.

Diseases of the Digestive System

The organs that form the digestive tract are the mouth, pharynx, oesophagus, stomach, intestines and the annexed glands, viz.: the salivary, liver, and pancreas. The development of these organs differs in the different species of animals. For example, so lipeds possess a small, simple stomach and capacious, complicated intestines.

Just the opposite is true of ruminants. The different species of ruminants possess a large, complicated stomach, and comparatively simple intestines. In swine we meet with a more highly developed stomach than that of so lipeds and a more simple intestinal tract. Of all domestic animals the most simple digestive tract occurs in the dog. These variations in the development of the different organs of digestion, together with the difference in the character of the feed and method of feeding, cause a variation in the kind of diseases met with in the different species. The complicated stomach of ruminants predispose them to diseases of this portion of the digestive tract. Because of their complicated intestinal tract so lipeds are prone to intestinal disease.

Diseases of the Mouth

General Discussion : The mouth is the first division of the digestivetr act. It is formed by the lips, cheeks, palate, soft palate, tongue and teeth. Here the feed is acted on mechanically. It is broken up by the teeth and moved about until mixed with the saliva and put into condition to pass through the pharynx and along the oesophagus to the stomach. The mechanical change that the feed is subject to is very imperfect in dogs. In the horse it is a slow, thorough process, although greedy feeders are not uncommon. The first mastication in the ox is three times quicker than in horses, but the process of rumination is slow and thorough.

Stomatitis

Simple inflammation of the mouth is frequently met with in horses. Ulcerative or infectious inflammation commonly occurs in young, and occasionally in old, debilitated animals. This form of sore mouth will be discussed along with other infectious diseases, and the following discussion will be confined to the non-infectious form of the disease. The causes are irritation from the bit, sharp teeth, irritating drenches, roughage that contains beards or awns of grasses and grains, and burrs that wound the lining membrane of the mouth. Febrile, or digestive disorders, or any condition that may interfere with feeding, may cause this disorder. In the latter cases the mucous membrane of the mouth is not cleansed by the saliva. Particles of feed may decompose and irritating organisms set up an inflammation. Putrid or decomposed slops, hot feeds, irritating drenches and drinking from filthy wallows are common causes of inflammation of the mouth in hogs. The symptoms vary in the different cases and species.

Slight or localized inflammation of the mouth is usually overlooked by the attendant. Lampas of horses may be considered a local inflammation involving the palate. Lacerations of the cheek or tongue by the teeth, or irritating feed, usually result in a slight interference with prehension and mastication and more or less salivation. Salivation from this cause should not be confused with salivation resulting from feeding on white clover. In generalized inflammation of the mucous membrane, the first symptomusually noticed is the inability to eat. On examining the mouth we find the mucous membrane inflamed, hot and dry. A part may appear coated. In a short time the odor from the mouth is fetid. Following this dry stage of the inflammation is the period of salivation. Saliva dribbles from the mouth, and in severe cases it is mixed with white, stringy shreds of epithelium and tinged with blood.

In less acute forms of the disease, we may noticelittle blisters or vesicles scattered over the lining membrane of the lips, cheeks and tongue. The acute form of stomatitis runs a short course, usually a few days, and responds readily to treatment. Localized inflammation caused by irritation from teeth, or feeding irritating feeds, does not respond so readily to treatment. The treatment is largely preventive and consists largely in removing the cause. When the mouth is inflamed, roughage should be fed rather sparingly, and soft feeds such as slops, mashes, or gruels given in place of the regular diet. Plenty of clean drinking water should be provided. In the way of medicinal treatment antiseptic and astringent washes are indicated. A four per cent water solution of boric acid may be used, or a one-half percent water solution of a high grade coal-tar disinfectant. The mouth should be thoroughly irrigated twice daily until the mucous surfaces appear normal.

Depraved Appetite

A depraved appetite is met with in all species of farm animals, but it is especially common in ruminants. It should not be classed as a disease, but more correctly as a bad habit, or symptom of innutrition or indigestion. The animals affected seem to have an irresistible desire to lick, chew and swallow indigestible and disgusting objects.

The common cause of depraved appetite is the feeding of a ration deficient in certain food elements. A ration deficient in protein or in salts is said to cause this disorder. Lack of exercise, or confinement, innutrition, and a depraved sense of taste may favour the development of this disease. For example, when sheep are housed closely they may contract the habit of chewing one another's fleeces. Lambs are especially apt to contract this habit when suckling ewes that have on their udders long wool soiled with urine and faeces.

The first symptom is the desire to chew, lick or eat indigestible or filthy substances. Horses and cattle may stand and lick a board for an houror more; cattle may chew the long hair from the tails of horses; sheep may nibble wool; sows may within a short time after giving birth to their pigs, kill and eat them; chickens may pick and eat feathers. Innutrition may accompany the abnormal appetite, as very frequently the affected animal shows a disposition to leave its feed in order to eat these injurious and in nutritions substances. In ruminants, the wool or hair may form balls and obstruct the opening into the third compartment, causing chronicin digestion and death.

The treatment consists in the removal of the cause. Feeding a ration that meets the needs of the system, clean quarters and plenty of exercise are the most important preventive lines of treatment. In such cases medicinal treatment (saline and bitter tonics) may be indicated. It is usually advisable to remove the affected animals from the herd or flock in order to prevent others from imitating them.

Diseases of the Stomach

There is a remarkable difference in the development of the stomachs of so lipeds and ruminants. The horse's stomach is simple and has a capacity of three or four gallons. The left portion is lined with a cuticular mucous membrane, and the right portion with a glandular mucous membrane that has in it the glands that secrete the gastric juice. The most important digestive change in the feed is the action of the gastric juice on the proteids and their conversion into the simpler products, proteoses and peptones.

Ruminants have a compound stomach : The capacity of the stomach of the ox is between twenty and thirty gallons. The four compartments into which it is divided are the rumen, reticulum, omasum, and abomasum or true stomach. The rumen is the largest compartment, with a capacity of more than twenty gallons. The reticulum is the smallest, with a capacity of about one-half gallon.

After a brief mastication, the food passes directly to the rumen . Here it is subjected to a churning movement that mixes and presses the contents of the rumen forward in the direction of the oesophageal opening, where it is ready for regurgitation. It is then carried back to the mouth, remasticated and returned to the rumen. This is termed rumination. All food material that is sufficiently broken up is directed toward the opening into the third compartment by the oesophageal grove, a demi-canal that connects this with the oesophageal opening.

The third compartment, the omasum, communicates anteriorly with the second and first, and posteriorly with the fourth compartment or true stomach. The interior arrangement of this compartment is most singular. It is divided by a number of large folds of the lining membrane between which are smaller folds. It is between these folds that the contents pass.

The first three compartments possess no glands capable of secreting adigestive juice. However, important digestive changes occur. The carbohydrates are digested by means of enzymes contained in the feed. The most important function of the rumen and omasum is the maceration of the fibrous substances, and the digestion of the cellulose. Between sixty and seventy per cent of the cellulose is digested in the rumen.

The abomasum is lined by a gastric mucous membrane. The gastric juice secreted converts the protein into peptones. In the young a milk curdling ferment is also secreted by the glands of this compartment. The Stomach Of The Hog is a type between the carnivora and ruminant. The digestive changes may be divided into four stages. The first period is one of starch conversion; the second period is the same, only more pronounced; the third period, both starch and protein conversion occurs; and the fourth period is taken up mostly with protein digestion.

Acute Indigestion of the Stomach of So lipeds

Diseases of the stomach areless common in so lipeds than in ruminants. The simple stomach of the horseand the comparatively unimportant place that it occupies in the digestion of the feed renders it less subject to disease. Only under the most unfavourable conditions for digestion of the feed does this class of disorders occur. Acute indigestion in the form of overloading and fermentation occurs in the stomach.

The predisposing causes that have to do with the development of these disorders, are the small capacity of the stomach and the location and smallness of the openings leading from the oesophagus and into the small intestines. Greedy eaters are more prone to indigestion than animals that eat slowly and are fed intelligently.

The following exciting causes may be mentioned: Sudden changes in ration; feeding too much green feed or grain; feeding frozen or decomposed feeds; drinking ice-cold water; and violent exercise or work that the animal is not accustomed to, immediately after feeding are the common disease-producing factors.

The symptoms may vary from impaired appetite and slight restlessness to violent, colicky pains. In the large majority of cases the attendant is unable to differentiate between this and other forms of acute indigestion. The characteristic symptoms are attempts at regurgitation and vomiting, assuming a dog-sitting position and finally such nervous symptoms as champing of the jaws, staggering movement and extreme dulness.

The violent form of gastric indigestion frequently ends in death. Rupture of the stomach is not an uncommon complication.

The treatment is both preventive and medicinal. This digestive disorder can be prevented. The feeding of the right kind of a ration and in the right way, and avoiding conditions that may interfere with the digestion of the feed, are the general lines of preventive treatment indicated. Such measures are of special importance in the handling of animals that possess an individual predisposition toward this class of disease. In mild attacks the animal should be subjected to a rigid or careful diet during the attack and for a few days later.

It is advisable to place the animal in a comfortable stall that is well bedded with straw and plenty large for it to move about in. If a roomy sick-stall can not be provided, a grass lot or barn floor may be used. If the weather is chilly or cold, the body should be covered with a blanket and roller bandages applied to the limbs.

Bulky drenches should not be given. Stimulants and drugs capable of retarding fermentation are indicated. Sometimes the administration of a sedative is indicated. Treatment should be prompt, as in many cases fermentation of the contents of the stomach occurs and gases form rapidly. From two to four ounces of oil of turpentine may be given in from six to eight ounces of linseed oil.

Acute Indigestion of the Stomach of Ruminants

The different forms of acute indigestion are bloating, overloading of the rumen and impaction of the omasum.

Tympanites, "Bloating"

This disorder is usually caused by animals feeding on green feeds, such as clover, alfalfa and green corn, that ferment readily. Stormy, rainy weather seems to favour bloating. The consumption of spoiled feeds such as potatoes and beets may cause it. The drinking of a large quantity of water, especially if cold, chills the wall of the rumen and interferes with its movement. Frozen feeds may act in the same way. Sudden changes in the feed, inflammation of the rumen, and a weak peristaltic movement of the paunch resulting from disease or insufficient nourishment are frequent causes. It may occur in chronic disease. In tuberculosis, bloating sometimes occurs.

The symptoms are as follows: The paunch or rumen occupies the left side of the abdominal cavity, hence the distention of the abdominal wall by the collecting of gas in the rumen occurs principally on that side. The gas forms quickly and the distended wall is highly elastic and resonant. The animal stops eating and ruminating, the back may be arched and the ears droop. In the more severe cases the wall of the abdomen is distended on both sides, the respirations are quickened and laboured, the pulse small and quick, the eyes are prominent and the mucous membrane congested. Death results from asphyxia brought on by the distended paunch pushing forward and interfering with the movement of the lungs and the absorption of the poisonous gases.

The treatment is both preventive and medicinal. This form of acute indigestion can be largely prevented by practising the following preventive measures: All changes in the feed should be made gradually, especially if the ration fed is heavy, or the new ration consists largely of green, succulent feed. Cattle pasturing on clover should be kept under close observation. It is not advisable to pasture cattle on rank growths of clover that are wet with dew or a light rain. Bloating can be quickly relieved by puncturing the wall of the paunch with the trocar and cannula. The operation is quite simple and is not followed by bad results. The instrument is plunged through the walls of the abdomen and rumen in the most prominent portion of the flank, midway between the border of the last rib and the point of the haunch. The trocar is then withdrawn from the cannula. After the gas has escaped through the cannula, the trocaris replaced and the instrument withdrawn. After using the trocar and cannula, the instrument should be cleaned by placing it in boiling hot water. It is advisable to wash the skin at the seat of the operation with a disinfectant before operating. In chronic tympanitis, it is sometimes advisable to leave the cannula in position by tying a tape to the flange, passing it around the body and tying.

As a cathartic for cattle, we may give one quart of linseed and from two to four ounces of turpentine, or one to two pounds of Epsom or Glauber's salts, dissolved in plenty of water. Sheep may be given about one-fourth the dose recommended for cattle.

Overloading the Rumen

This form of indigestion occurs when ruminants have access to feeds that they are not accustomed to. As a result, they eat greedily and the mass of feed in the rumen becomes so heavy that the walls of the organ can not move it about, and digestion is interfered with. This is especially true of succulent feeds. A diseased condition of the animal predisposes it to this disorder. If after eating an excessive amount of dry, in nutritions fodder, the animal drinks freely of cold water, acute symptoms of overloading are manifested.

The general symptoms occurring in overloading resembles those seen in bloating. The symptoms may be mild and extend over a period of several days, or it may take on a highly acute form, terminating fatally within a few hours. The acuteness of the attack depends on the character and quantity of feed eaten. If a large quantity of green feed is eaten, fermentation occurs and the animal may die within a few hours. The swelling on the left side has a doughy feel. It is not as elastic and resonant as in bloat, even when complicated by some gas formation. The animal may stop ruminating, refuse to eat, and act dull. In the more severe cases the respirations are hurried and laboured, the pulse small and quick and the expression of the face indicates pain. Colicky pains sometimes occur. Death may occur from shock or asphyxia.

The treatment is both preventive and curative. This disease can be prevented by using the necessary precautions to prevent animals from overeating. If gas forms, the trocar and cannula should be used. A drench of from one to two pounds of Epsom or Glauber's salts should be given. Sheep may be given from four to six ounces of Epsom or Glauber's salts. Weshould endeavour to stimulate the movement of the paunch by pressure on the flank with the hand, throwing cold water on the wall of the abdomen and by hypodermic injections of strychnine. Rumenotomy should be performed when necessary. This operation consists in opening the walls of the abdomen and rumen, and removing a part of the contents of the rumen. This is not adangerous operation when properly performed, and should not be postponed until the animal is too weak to make a recovery.

Impaction of the Omasum

This disease may occur as a complication of other forms of acute indigestion and diseases accompanied by an abnormal body temperature. Feeds that are dry and in nutritions commonly cause it. Other causes are an excessive quantity of feed, sudden changes in the diet and drinking an insufficient quantity of water.

As in other diseases of the stomach, the appetite is diminished, rumination ceases or occurs at irregular intervals, and the animal is more or less feverish. Bloating and constipation may occur. The animal may lose flesh, is weak, walks stiffly and grunts as though in pain when it moves about in the stall and at each respiration. In the acute form, marked symptoms are sometimes manifested. At first the animal acts drowsy; later violent nervous symptoms may develop.

The course of this disease varies from a few days to several weeks. Death frequently occurs. Frequently a diarrhoea accompanies recovery, a portion of the faeces appearing black with polished surfaces, as though they had been baked. The preventive treatment consists in practising the necessary precautions against the development of this disease by avoiding sudden changes in the feed, the feeding of dry, in nutritions feeds in too large amounts, allowing animals plenty of water and providing them with salt.

The best purgative to give is Glauber's or Epsom salts in from one-to two-pound doses, dissolved in at least one gallon of water. This physic may be repeated in from twelve to eighteen hours if necessary. Two drachms of tincture of nux vomica and one ounce of alcohol may be given in a drench three times daily. Hypodermic injections of strychnine, eserine, or pilocarpine are useful in the treatment of this disease. When recovery begins, the animal should be allowed moderate exercise and be fed food of a laxative nature.

Foreign Bodies in the Stomach of Ruminants

Foreign bodies such as hair balls and wire are very commonly found in the reticulum. This is because of the habits of this class of animals. Cattle eat their feed hastily and do not pick it over as carefully as does the horse.

Smooth, round objects do no appreciable harm unless they block the opening into the third compartment of the stomach. This frequently occurs in wool-eating lambs. Sharp-pointed objects may penetrate the surrounding tissues or such organs as the spleen, diaphragm, and pericardial sack. If these organs are injured by the foreign body serious symptoms develop. The general symptoms are pain, fever, weakness and marked emaciation. It is very difficult to form a correct diagnosis, as the disease comes on without any apparent cause. Sometimes a swelling is noticed in the right and inferior abdominal region. If the heart becomes injured, symptoms of pericarditis are manifested.

The treatment is largely preventive. Special care should be used to avoid getting foreign substances into the feed given to cattle. The feed troughs should be kept clean; we should avoid dropping nails and staples into the feed when repairing the silo or grain bin; and pieces of baling wire should be removed from straw or hay. Feeds known to be dirty should be run through a fanning mill before feeding.

Inflammation of the Stomach of Swine

Overloading and feeding spoiled feed are common causes of inflammation of the stomach. Swill-fed hogs are most commonly affected with this disorder. Overloading more often results in an inflammation of the stomach if the overloading follows the feeding of alight ration, and the weather is extremely warm. Hogs that are accustomed to eating salt may eat too much of it when fed to them after it is with held for a week or longer, and a large quantity of water is taken soon afterwards. Slop containing alkaline washing powders and soaps irritate the stomach and intestines and cause a serious inflammation.

The symptoms are loss of appetite, restlessness and sometimes colicky pains. The hog usually wanders off by itself, acts dull, grunts, lies down in a quiet place or stands with the back arched and the abdomen held tense. Vomiting commonly occurs. Sometimes the animal has a diarrhoea. The body temperature may be above normal.

The treatment consists in avoiding irritating feeds and sudden changes in the kind or quantity of feed fed. Drenching with hot water, or with about one ounce of ipecacuan may be practised. From one to three ounces of castoroil, depending on the size of the hog, may be given. After recovery the hogs should be confined in a comfortable pen and fed an easily digested ration.

Diseases of the Intestines

General Discussion

The intestinal tract of so lipeds is the best developed of any of the domestic animals. It is divided into two portions, small and large. The small intestine is a little over seventy feet in length and about one and one-half inches in diameter. The mucous membrane lining presents a large absorbing surface and is well supplied with absorbing vessels that take up the sugars, proteids and fats, which are finally distributed to the body cells by the blood capillaries. In addition to these absorbing vessels the mucous membrane contains intestinal glands that secrete the intestinal juice. Other digestive secretions from the pancreatic gland and the liver are poured into the small intestine near its origin. These digestive juices act on the proteids, sugars, starches and fats, changing them into substances that are capable of being absorbed.

After disengaging itself from the mass of loops lodged in the region of the left flank, the small intestine crosses to the region of the right flank, where it terminates in the first division of the large intestine.

The large intestine is formed by the following divisions: caecum, double colon, floating colon and rectum. The caecum is a large blind pouch that has a capacity of about seven gallons. The double colon is the largest division of the intestines. It is about twelve feet in length and has a capacity of about eighteen gallons. This portion of the intestine terminates in the region of the left flank in the floating colon. The latter is about ten feet in length and about twice the diameter of the small intestine, from which it can readily be distinguished by its sacculated walls. The rectum is the terminal portion of the intestinal tract. It is about one and one-half feet in length and possesses heavy, elastic walls.

Fermentation and cellulose digestion occur in the caecum and double colon. It is in the floating colon that the faeces are moulded into balls. The faeces are retained in the rectum until defecation takes place. The intestinal tract of cattle is longer than that of so lipeds and the different divisions are not as well defined as in the horse's intestine and about one-half its diameter. The large intestine is about thirty-five feet in length and its capacity six or seven gallons.

Acute Intestinal Indigestion of So lipeds

Acute indigestion is more common in horses and mules than it is in any of the other domestic animals. Because of the difference in the causes and symptoms manifested, we may divide it into the following forms: spasmodic, flatulent and obstruction colic.

The predisposing causes are general and digestive debility resulting from the feeding of an insufficient or unsuitable ration, and general and parasitic diseases of the intestine. Nervous, well-bred horses are most susceptible to nervous or spasmodic colic.

The direct causes are improper methods of feeding and watering; giving the animal severe or unusual exercise immediately before or after feeding; the feeding of spoiled or green feeds and new grains; chilling of the body; imperfect mastication of feed because of defective teeth; obstruction of the intestine by worms.

The feeding of grain at a time when the animal is not in fit condition to digest it results in imperfect digestion in both the stomach and intestine. This leads to irritation of the intestine and abnormal fermentation of its contents. The drinking of a large quantity of water immediately after feeding grain flushes at least a part of the undigested grain from the stomach through the small intestine and into the caecum. New grains, such as new oats, are hurried along the small intestine and reach the large intestine practically undigested. The two latter conditions are common causes of flatulent or wind colic. Sudden change in the ration, especially to a green feed, may result in intestinal irritation and flatulence.

Horses that are greedy feeders and have sharp, uneven, smooth or diseased teeth are unable to masticate the feed properly. This results in unthriftiness caused by imperfect digestion and assimilation of the feed. Such animals usually suffer from a catarrhal or chronic inflammation of the intestine, and may have periodic attacks of acute indigestion or colic.

Obstruction colic is very often caused by the feeding of too much roughage in the form of straw, shredded fodder, or hay. Debility of ten contributes to this form of indigestion, and the double colon may become badly impacted with alimentary matter.

Worms may irritate the intestinal mucous membrane and interfere with digestion, obstruct the intestine and cause debility and circulatory disturbances. The large round worm may form a tangled mass and completely fill a portion of the double colon.

Some species attach themselves to the intestinal wall, suck the blood of the host and cause anaemia and debility. Colic resulting from circulatory disturbances is not common. The female of a certain species of strongulus deposits eggs in the mucous membrane. On hatching, the larvae may enter a blood capillary, drift along in the blood stream and finally come to rest in a large blood-vessel that supplies a certain portion of the intestines with blood. Here the parasite develops. The wall of the vessel becomes irritated and inflamed, pieces of fibrin flake off and drift along the blood stream until finally a vessel too small for the floating particle to pass through is reached and the vessel becomes plugged. The loop of intestine supplied by it receives no blood. A temporary paralysis of the loop occurs, which persists until a second vessel is able to take over the function of the one that is plugged. This form of colic is most common in old horses.

Such complications of acute indigestion as twisting, in folding and displacement of the intestine may occur. It is not uncommon for a stallion to suffer from strangulated hernia, due to a rather large in ternalinguinal ring and a loop of the intestine passing through it and into the inguinal canal or scrotum. Such displacements are usually accompanied by severe colicky pains.

The symptoms vary in the different cases. In the mild form, the colicky pains are not prominent, but in the acute form, the animal is restless, getting up and down in the stall and rolling over. These movements are especially marked when the abdominal pain is severe.

In the spasmodic form the attack comes on suddenly, the colicky pains are severe, and the peristaltic movement of the intestine is marked and accompanied by loud intestinal sounds. In most cases of indigestion characterized by fermentation and collections of gas in the intestine there is gastric tympany as well.

Acute indigestion characterized by impaction of the large intestine pursues a longer course than the forms just mentioned, and the abdominal pain is not severe.

Congestion and inflammation of the intestine may result from the irritation produced by the feed. When this occurs, the abdominal pain is less violent. The animal usually acts dull, the walk is slow and unsteady, and the respirations and pulse beats may be quickened.

A large percentage of the cases of acute indigestion terminate fatally. The course of the disease varies from a few hours to several days.

The treatment is both preventive and curative. The preventive treatment is by far the most important. This consists in observing right methods of feeding and caring for horses. The attendant should note the condition of the animal before feeding grain, feed regularly and avoid sudden changes in feed. If a horse has received unusual exercise, it is proper to feed hay first, and when the animal is cooled out, water and feed grain. Drinking a small quantity of water when tired or following a meal is not injurious, but a large quantity of water taken at such times is injurious and dangerous to the health of the animal. The feeding of spoiled or mouldy feeds to horses is highly injurious.

The horse should be given a roomy, comfortable stall that is well bedded, or a clean grass lot. If the attack appears when the animal is in harness, we should stop working it and remove the harness immediately. Work or exercise usually aggravates the case and may cause congestion and inflammation of important body organs. In cold weather the animal should be protected by blankets. If the pain is violent, sedatives may be given. The gaseous disturbances should be relieved by puncturing the wall of the intestine with the trocar and cannula. Rectal injections of cold water maybe resorted to. Fluid extract of cannabis indica in quarter ounce doses and repeated in one hour may be given in linseed oil. In all cases it is advisable to drench the animal with one pint of raw linseed oil and two ounces of turpentine. Strychnine, eserine and pilocarpine are the drugs commonly used by the veterinarians in the treatment of acute indigestion. Small and repeated doses of the above drugs are preferred to large doses. This is one of the diseases that requires prompt and skilled attention.

Sharp, uneven or diseased teeth should receive the necessary attention. In old horses, chopped hay or ground feeds should be fed when necessary. Debility resulting from hard work, wrong methods of feeding and intestinal disorders must be corrected before the periodic attacks of indigestion can be relieved. If the presence of intestinal worms is suspected, the necessary treatment for ridding the animal of these parasites should be resorted to.

Bitter or saline tonics should be administered in the feed when necessary. The following formula is useful as a digestive tonic: Sodium bicarbonate and sodium sulfate, one pound of each, powdered gentian one-half pound, and oil meal five pounds. A small handful of this mixture may be given with the feed two or three times daily.

Inflammation of the Intestines

The same causes mentioned in inflammation of the stomach and acute indigestion may cause this disease. It is most frequent at times when there are great variations in the temperature. Sudden cold or any influence that chills the surface of the body, orinternal cold caused by drinking ice water or eating frozen feed, may cause it. The infectious forms of enteritis are caused by germs and ptomaines in the feed. Drinking filthy water or eating spoiled, mouldy feeds are common causes. In cattle pasturing in low, marshy places, enteritis may be common. The toxic form is caused by irritating poisons, such as caustic acids, alkalies and meat brine.

In the mild form of enteritis the appetite is irregular, the animal acts dull and stupid and may be noticed lying down more than common. Slight abdominal pains occur, especially following a meal. An elevation in the body temperature may be noted and the animal may drink more water than usual. Constipation or a slight diarrhoea may be present. The feces may be soft and foul smelling, coated with mucus, and slightly discoloured with blood.

In the severe form of enteritis pressure on the abdomen may cause pain, the respiration and pulse beats are quickened and the body temperature is elevated. The abdominal pain may be severe and the animal is greatly depressed or acts dull. The movement of the intestines is suppressed at first and constipation occurs. Fermentation and the formation of gas may take place. Later the intestinal peristalsis increases and a foul-smelling diarrhoea sets in that is often mixed with blood. In the toxic form there may be marked nervous symptoms. Spasms, convulsions, stupefaction and comamay be manifested. In the mild form recovery usually occurs within a few days. The more serious forms of the disease do not terminate so favourably. In the toxic form death usually occurs within a few days.

The large majority of cases of enteritis can be prevented by practising the necessary preventive measures. It is very necessary that animals exposed to cold be provided with dry sleeping quarters that are free from draughts. Where a number of animals are fed a heavy grain ration, or fed from the same trough, they should be kept under close observation. This is necessary in order to detect cases of indigestion or over feeding early, and resort to the necessary lines of treatment, so as to prevent further irritation to the intestinal tract. Live stock should not be forced to drink water that is ice-cold. Low, poorly-drained land is not a safe pasture for cattle and horses. Spoiled roots, grains and silage, mouldy, dirty roughage and decomposed slops should not be fed to live stock.

The treatment consists in withholding all feed and giving the animal comfortable, quiet quarters—warm quarters and protection from the cold, providing the animal with a heavy straw bed, or with blankets if necessary, if the weather is cold. From five to forty grains of calomel may be given, depending upon the size of the animal and the frequency of the dose, two or three times a day. In case the animal is suffering severe pain, morphine given hypodermically may be indicated. In the mild form and at the very beginning of the attack, linseed oil may be administered to the larger animals. The dose is about one quart. The smaller animals may be given castor oil in from one-to four-ounce doses.

When convalescence is reached the animal should be fed very carefully, as the digestive tract is not in condition to digest heavy rations or feeds that ferment readily.

Diarrhoea

Diarrhoea occurs as a symptom of irritation and inflammation of the intestinal mucous membrane. Sudden changes in the feed, the feeding of a succulent green ration, severe exercise when the animal is not in condition for it, and chronic indigestion may cause diarrhoea in the absence of an intestinal inflammation.

The following symptoms may be noted: Animals affected by a diarrhoea act dull and weak; thirst is increased and the animal may show evidence of fever; the intestinal evacuations are soft, thin, and sometimes have an offensive odor. If the diarrhoea continues for several days, the animal loses flesh rapidly and the appetite is irregular. In such cases weakness is a prominent symptom.

Recovery usually occurs when the animal is dieted and rested. The treatment consists in giving a physic of linseed or castor oil. Horses and cattle may be given from one-half to one quart of linseed oil; sheep and hogs from one to four ounces of castor oil. Feed should be with held. Morphine may be given hypodermically to the large animals after a period of six to eight hours following the administration of the physic.

The following formula is quite useful in checking diarrhoea: salol one-halfounce, bismuth sub nitrate one ounce, and bicarbonate of soda two ounces. The dose of this mixture is from one to four drachms, depending on the size of the animal, three or four times a day.

White Scours or Diarrhoea in Young Animals

Young animals, when nursing the mother or fed by hand, frequently develop congestion and inflammation of the stomach and intestines. This disorder is characterized by adiarrhoea.

The causes may be grouped under two heads: wrong methods of feeding and care, and specific infection.

The first milk of the mother is a natural laxative and aids in ridding the intestine of the young of such waste material (meconium) as collects during fetal life. If this milk is withheld, the intestine becomes irritated, constipation occurs, followed by a diarrhoea or serious symptoms of a nervous character, caused by the poisonous effect of the toxic substances absorbed from the intestine on the nervous system.

Changes in the ration fed the mother, excitement, unusual exercise and disease change the composition of the mother's milk. Such milk is irritating to the stomach and intestines of the young. This irritation does not always develop into a diarrhoea, but may result in a congestion of the stomach.

When the young are raised artificially or by hand, and fed milk from different mothers of the same or different species, or changed from whole to skim milk, acute and chronic digestive disorders that are accompanied by a diarrhoea are common. Feeding calves from filthy pails, allowing them to drink too rapidly and giving them fermented milk are common causes of scours. White scours caused by irritating germs is a highly infectious disease. The disease-producing germs gain entrance to the body by way of the digestive tract and the umbilical cord.

Insanitary conditions, such as dark, cold, damp, filthy quarters, lower the vitality of young animals, and predispose them to digestive disorders as well as other diseases.

The symptoms are as follows: Constipation accompanied by a feverish condition precedes the diarrhoea; colicky pains are sometimes manifested; the diarrhoea is usually accompanied by depression, falling off in appetite and weakness. At first the intestinal discharges are not very foul smelling; later the odor is very disagreeable. The faeces may be made up largely of undigested, decomposed milk that adheres to the tail and hind parts. If the diarrhoea is severe, the animal refuses to suckle or drink from the pail, and loses flesh rapidly. It is usually found lying down. The ears droop and the depression is marked. The body temperature may vary from several degrees above to below the average normal.

The infectious form of white scours may be diagnosed by the history of the outbreak. In this form of the disease, a large percentage of the young are affected and the death-rate is very high.

Calves and lambs frequently die of an acute congestion of the fourth stomach. In this disease, the symptoms appear shortly after feeding. It is characterized by colicky pains, convulsions and coma. The treatment is largely preventive. Young animals should be provided with dry, clean, well-ventilated quarters and allowed plenty of exercise. Colts thrive best if allowed to run in a blue grass pasture with the mother. If the mother is worked, suitable provisions in the way of quarters and frequent nursing should be provided. Calves, lambs and pigs are the most frequent sufferers from insanitary quarters. In breeding, we should always strive to get strong, vigorous, healthy young. The care given the mother in the way of exercise and feeding is an important factor here.

The first milk of the mother should not be withheld from the young, especially if the animal is raised by hand. We must also feed it regularly and not too much at any one time. Any change in the milk should be made gradually, and it is usually advisable to reduce the ration slightly when such a change is made, so as not to over work the digestive organs. Pails and bottles from which the animal feeds should be kept clean.

Colts raised on cow's milk must be fed and cared for carefully. The milk must be sweet and made more digestible by diluting it with one-third water. A little sugar may be added. It is very advisable to add from one-half to one ounce of lime water to each pint of milk fed. Frequent feeding is very necessary at first, and we must not underestimate the quantity of milk necessary to keep the colt in good condition. It should be taught to eat grain as soon as possible.

Because of the irritated condition of the stomach and intestine, the animal suffering from diarrhoea is unable to digest its feed. For this reason it is very important to withhold all feed for at least twelve hours. Water should be provided. The alimentary tract is relieved of the irritating material by giving the animal a physic of castor or linseed oil. The dose varies from one-quarter to one-half ounce for the lamb and from one to four ounces for the colt or calf. It is advisable in most cases to follow this with the following mixture: bicarbonate of soda one ounce, bismuth sub nitrate one-half ounce, and salol one-quarter ounce. The dose for the colt and calf is one teaspoonful three times a day. Lambs and pigs may be given from one-fourth to one-half the above dose. It is usually advisable to give ewes and sows a physic if their young develop a diarrhoea. Mothers that are heavy milkers may be given a physic the second or third day following birth. The ration should be reduced as well during the first week.

Diseases of the Digestive Tract of Poultry

General Discussion

The digestive tract of poultry is composed of the following organs: mouth, gullet, crop, stomach, gizzard and intestines, with the two large glands, the liver and pancreas. The digestion of the feed begins in the crop. Here the feed is held for a short time, mixed with certain fluids and softened. On reaching the stomach it becomes mixed with the digestive fluid secreted by the gastric glands. This second digestive action consists in thoroughly soaking the feed in the gastric juice, making it soft and preparing it for maceration by the heavily muscled gizzard. Following maceration it passes into the intestine. It is here that the digestive action is completed and absorption occurs.

Under the conditions of domestication, poultry are subject to a great variety of intestinal disorders.

Diseases of the Crop

Impaction and inflammation are the two common diseases of the crop. Large, impacted crops are usually caused by the feeding of too much dry feed, fermentation of the contents of the crop and foreign bodies that obstruct the opening from the organ.

Inflammation of the crop is caused by excessive use of condiments in the feed, putrid or spoiled feeds and eating caustic drugs, such as lime and rat poison.

The symptoms are dulness, an indisposition to move about, drooping wings and efforts to eject gases and liquids. The crop is found greatly distended and either hard or soft, depending on the quantity of feed present and the cause of the distention. If fermentation is present the crop usually feels soft.

The preventive treatment consists in practising proper methods of feeding. The curative treatment of a recent case consists in manipulating the mass of feed, breaking it up and forcing it upwards toward the mouth. If difficulty in breaking up the mass is experienced, it is advisable to administer a tablespoonful of castor oil to the bird.

If the above manipulations are unsuccessful, an operation is necessary. This consists in making an opening through the skin and the wall of the crop and removing the contents with tweezers. The opening must be closed with sutures. The proper aseptic precautions must be observed. In inflammation of the crop, the bird should be dieted for at least one day, and one teaspoonful of castor oil given as a laxative.

Acute and Chronic Indigestion

The recognition of special forms of indigestion in poultry is difficult. A flock of poultry that is subject to careless and indifferent care may not thrive and a number of the birds develop digestive disorders. This may be indicated by an abnormal or depraved appetite and emaciated condition. Constipation or diarrhoea may occur. In the more severe cases the bird acts dull, the feathers are ruffled and it moves about very little. The treatment consists in removing the cause, and giving the flock atonic mixture in the feed. The following mixture may be used: powdered gentian and powdered ginger, eight ounces of each, Glauber's salts four ounces, and sulfate of iron two ounces. One ounce of the above mixture maybe given in ten pounds of feed.

White Diarrhoea of Young Chickens

White diarrhoea is of the greatest economic importance to the poultryman. The loss of chicks from this disease is greater than the combined loss resulting from all other diseases. It is stated by some authors that not less than fifty per cent of the chicken shatched die from white diarrhoea.

Such a heavy death-rate as is attributed to this disease can not result from improper methods of handling and insanitary conditions. Before it was proven that white diarrhoea was caused by specific germs, a great deal of emphasis was placed on such causes as debilitated breeding stock, improper incubation, poorly ventilated, overcrowded brooders, too high or too low temperatures and filth. Such conditions are important predisposing factors, and may, in isolated cases, result in serious intestinal disorders.

The microorganisms causing this disease belong to both the plant and animal kingdoms. Infection usually occurs within a day or two following hatching. Chicks two or three weeks of age seldom develop the acute form of the disease. Incubator chicks are the most susceptible to the disorder.

The following symptoms occur: The chicks present a droopy, sleepy appearance; the eyes are closed, and the chicks huddle together and peep much of the time; the whitish intestinal discharge is noticed adhering to the fluff near the margins of the vent, and the young bird is very weak; death may occur within the first few days. After the first two weeks the disease becomes less acute. In the highly acute form the chicks die without showing the usual train of symptoms. It is very easy to differentiate between the infectious and the non-infectious diarrhoea. In the latter, the percentage of chicks affected is small and the disease responds to treatment more readily than does the infectious form. The death-rate in the latter form is about eighty percent.

The treatment of diarrhoea in chicks from any cause is preventive. This consists in removing the cause. No person can successfully handle poultry if he does not give the necessary attention to sanitation. Poultry houses, runs, watering fountains and feeding places must be constantly cleaned and disinfected. The degree of attention necessary depends on the surroundings, the crowded condition of the poultry houses and runs, and the presence of disease in the flock. If disease is present, we can not clean and disinfect the quarters too often. The attendant can not overlook details in handling the incubator or brooder and feeding the chicks and be uniformly successful.

If the disease is known to be present in the flock, the incubators and brooders should be thoroughly disinfected by fumigating them with formaldehyde gas. If dirty, they should first be washed with a water solution of a good disinfectant. For a period of from twenty-four to forty-eight hours after hatching, the chicks should receive no feed. Dr. Kaupp recommends as an intestinal antiseptic, sulfocarbolate thirty grains, bichloride of mercury six grains, and citric acid three grains, dissolved in one gallon of water. This solution should be kept in front of the chick sall the time. A water solution of powdered copper sulfate (about one-halfteaspoonful dissolved in one gallon of water) may be used.

12

Landless Livestock Farming

Goat Production by Landless and Small Scale Farmers in North Indian Plains

Goat production has flourished in poorly endowed areas like dry zones and the mountainous areas in developing countries of Asia and Africa where over 95% of the world goat population is concentrated. Goats are reared mainly by poorest of the poor of the poor developing countries. In India too, the small and marginal farmers, including landless agricultural labourers, mostly in non green revolution areas where irrigation facilities are poorly developed, prominently rear goats (Rekib, 1998). Many macro level studies have confirmed that goats are important for the people living in dry areas and mountainous regions. Also, it is now well documented that goat production can be an important tool to ameliorate the conditions of poor women in developing countries, thus, being seen as source of women's emancipation by the development agencies.

Goat population witnessed +8.1% change between 1990 and 1996 at a global level and the maximum change (+27%) has been in Asian countries. Over the last 15 years, the number of goats has increased by almost 50% at world level, whereas, sheep decreased by 4% and cattle increased by only 9%. Thus, goat has emerged as a major livestock species that is enormously rising in number.

In India, during the last 40 years, the goat population rose by 140%. Goat population in India has increased at a faster rate than that of other species of farm animals. In 1952, there were 47.0 million goats, which has risen now to 123.0 million (17.08 % of the world's goat population) making India number two in the world ranking and number one in goat milk production. The National Commission on Agriculture (NCA, 1976) in India had recommended the reduction in goat population on the grounds of environmental implications of goat rearing on degradation of forest and grazing lands. The phenomenal increase in goat population despite NCA recommendations against it, is explained by the importance of goats for the rural people in India. In India, the contribution of livestock sector to Gross Domestic Product (GDP) is rising though the contribution of agricultural sector in general is witnessing a declining trend. In such a situation, goats are going to be even more important source of livelihood for many more people in coming years and, thus, deserve greater attention at both the micro and macro level. The current study being a part of a broader study on livestock production systems was undertaken mainly to appraise the role of goats for small scale farmers including the landless rural population in predominant crop-livestock mixed farming systems operating in India.

Methodology

The study was conducted in Bareilly district (Latitudes 28.10^{0}N and 28.50^{0}N and Longitudes 78.58^{0}E and 79.47^{0}E) of a north Indian State (Uttar Pradesh) during 1999-2001. The data were gathered on dynamics of goat production from primary as well as secondary sources. A door to door survey of 1721 households was carried out to find out the herd composition and association of livestock/goats owned with land holding, caste etc. in six selected villages. A Rapid Rural Appraisal (RRA) was conducted in two villages with 642 livestock owning households selected for more focused assessment of the situation.

Participatory Rural Appraisals (PRAs) were also undertaken with the family members of goat owning households so as to assess productivity, gender roles, disease patterns, marketing behaviour and distribution of goats along various dimensions especially caste, landholding and gender. Matrix ranking for choice of livestock species preferred by different categories of farmers, problem cause diagrams, constraints analysis, productivity potential of various species in different seasons, mortality/morbidity pattern, resource maps etc were developed using the RRA/PRA tools which further helped in appraisal of the goat production system in the area under study. However, the results are focused here only on association of goats with landholding and caste, which have significant implications for livestock production in India.

Result

Bareilly district is densely populated (688 persons/sq Km); 67% of the population is concentrated in rural areas involved mainly with agriculture and animal husbandry. The total livestock population in the district is 899,198, out of which 16,498 are crossbred cattle, 290,568 local cattle, 441,578 buffaloes, 4926 sheep, 116,076 are goats, 29,552 pigs. The average milk production per day for cow, buffalo and goat is 2.55, 3.35 and 0.5 Kgs, respectively. The low productivity may be explained in terms of low-input low-output system of livestock production prevailing by and large in India, low productivity potential of native breeds, acute shortage of feed and fodder, poor livestock health coverage, non commercial orientation of livestock keepers and poor marketing and extension support available to Indian farmers.

Uttar Pradesh with 13,110,000 goats (1992 Census), is the third largest state in India, after Rajasthan and Bihar, in terms of goat population. Whereas, goat is the third major livestock species (poultry excluded) after cattle and buffaloes reared in Bareilly district. There were 116,076 goats in the district. The total goat population across the six selected villages was 489. Goat command 3rd place (4th if poultry too is taken into account) in the study area, where, buffalo and cattle were having prominent positions.

The door-to-door survey of all the six selected villages was done to collect data on livestock profile. Out of the six villages, two villages namely Rithora and Khata were selected for more focused study on goat production system, since these two villages had maximum number of goats and highest concentration of goats, respectively. On the basis of the door to door survey, the general description of these villages in terms of livestock ownership in general. In Khata village for instance, all 54 goats were owned by the other backward castes (OBCs), while livestock owning families were only 1 and 4 from General and SC/ST categories, respectively.

In Rithora village, out of 244 goats, 223 were owned by OBCs, whereas, SC/ST (Socio economically downtrodden, constitutionally categorized and protected by the law) owned 15 goats and 6 goats were owned by farmers belonging to general caste category. The OBCs were the largest group having 441 families followed by SC/ST (92 families) and general caste (23 livestock owning families). The OBCs, which included Muslims, were the majority goat keepers across the selected villages; thus, goat is significant to them to supplement their family income. Besides, goats appeared to be very significant for the landless as they had maintained a maximum number of goats. There cannot be any better example of resourcelessness than being landless in a farming context, with limited opportunities of diversification for improving the quality of life. In the study area, such people have seen in goat a potential for supplementing their family income over their earnings from the wages as agricultural labourers.

Other general findings of the study can be summarized as follows:

1. Mostly Muslims and landless labourers kept goats. Muslims were included in the Other Backward Class (OBC) category (categorization according to economic status) which owned 426 goats (87.14%), out of the total 489 goats in the selected villages.
2. The farmers belonging to general castes* (Higher caste Hindus) in general, had less number of livestock and still less they had goats.
3. The OBCs among the Hindus in the area had least preferences for goats.
4. Small, marginal farmers and landless labourers were the group having interest in goat keeping. The area under forest was negligible in the district (313ha) as well as in the studied villages. Whereas, the proximity to forest positively affect goat rearing.

5. Goats were reared mainly for meat purpose and hectic activity was seen regarding the sale/purchase of goats during the Muslim festivals like Eid.
6. Goats are treated as ready cash as they could be disposed off easily, locally and at a reasonable price.
7. Women were taking more interest in nurture and even in the sale at home as well as at biweekly 'haat' (local market) at nearby town (Rithora), while children and aged women take goats to grazing.
8. Goats often cause conflict when these stray in the crop fields of other farmers. This is one of the deterrents for goat rearing especially for landless and small scale farmers.
9. The families which had more number of children were more inclined to keep goats as the children could be used for taking goats to grazing on fallow land, harvested fields, alongside roads and other uncultivated areas.

Breeding

Indiscriminate breeding is prevalent in the area. No efforts seem to have made to improve the stock through selective breeding or by introducing high potential breeds. The goats were small sized, colored (white/black/brown), showing no specific breed characters. The kidding takes place twice a year. One kid is usually born at I[st] kidding followed by twining in subsequent kidding. Goats of bigger body size are usually brought from outside (Etah, Etawah and Mathura) area for sale prior to Eid but the good stock is not maintained by the farmers for longer duration. This may be due to non availability of feed and fodder with the poor goat keepers. Male goats are taken better care of than the female ones as these fetch good price to the owners. Many do not like goat milk; hence, less priority is attached to milk traits of the goats.

Feeding

Feeding seems to be the most neglected aspect in goat production in the area under study. Goats were maintained on grazing in harvested fields, along the roadside and on other uncultivated/barren lands. The natural ability of goat to eat a wide variety of vegetation and waste has been, in fact, a big motivating factor for small, marginal and landless labourers to rear small number of goats. Stall-feeding in goats was very limited. She goats are given kitchen waste and the males are given inferior quality grains and grams for fattening purposes. Children and old members take goats for grazing in nearby fields in the morning and also in the afternoon. Those households, which do not have children or aged members in their families, hardly keep goats.

Disease Control

The farmers did not report disease in goats as problem. Since goat production is only a fringe activity for most of the farmers, the health problems of goats were hardly paid any attention. However, to find out the types of health problems in local goats, the information maintained in the register of Indian Veterinary Research Institute run weekly clinic at Rithora was also studied. This revealed that goats suffered mainly from worm load, mange, mineral deficiency, anorexia, contagious ecythyma, diarrhea, mastitis etc.

Management

Goats do not play a significant role in the day-to-day life of local farmers compared to the buffaloes and cattle. This may be due to the fact that the study area is well endowed with fertile soil, irrigation facilities, well connected by mettled road, inhabited by farmers belonging to Yadav and Kurmi castes who consider it lowly to rear goats. Moreover, the big majority of the farmers in the area were vegetarian. Yet, goats are considered by many as a ready cash in the times of immediate need/financial crisis, since, it find almost ready market all round the year. Goats are owned by and large by the small/marginal farmers and landless labourers, who owing to limitations of space and grazing area, can not maintain more than 3-4 goats at their homesteads. Housing and grazing seems to be a big constraint in goat husbandry for the local poor farmers. Its women mainly who take interest in goat rearing as the men generally concentrate on large animals.

Women find it a potential source of earning money exclusively for their personal needs as they could meet their personal expenses like in buying ornaments, cosmetics, toys and eatables for children etc with the occasional sale of goats. As such, they do take care of goats as their personal belonging or pocket money. Women especially from Muslim and socio economically weaker sections were seen increasingly in the by-weekly haat (local market), selling goats but not any other species, which were being sold by the men only. Elsewhere in the world too, women and children tend to be more caring for goats (Sinn; Ketzis and Chen, 1999). The bank officials in Rithora town informed that farmers never approached them for loans to buy goats, whereas, loans were sought for buying cattle and buffaloes. This indicates the possibility of suitable interventions including micro credit to women for goat husbandry that will lead to empowerment and emancipation of women.

Conclusion

Goat as a source of supplementing household income is getting increasing attention especially among the landless agricultural labourers and small and marginal farmers mostly belonging to Muslim and other marginalized groups. Women are increasingly finding it as a potential source of earning cash income to meet their personal requirements like buying ornaments, cloths etc. since the men mostly concentrate on large animals and pocket the earnings for other needs of family. The Muslim festival of Eid is one important factor for rearing goats in the area as a large number of goats are reared for sacrifice during this festival and rural people find it economically very rewarding to rear goats for sale.

The goat population is likely to get a further boost once the culinary preferences of other communities like Hindus who are in majority change in favour of non vegetarian foods. The goats of good productive potential are likely to be introduced in the area once the local demand rises owing to increasing awareness and industrialization. The landless, small and marginal farmers are gradually recognizing potential of goat as a low cost solution to resourcelessness. The large animals do not fit well into their household and infrastructural paradigms and goat has little requirements, thus, it is getting popular with advancing times.

Goats are a highly attractive asset for poor women in India as also in the area under study. Yet land, credit and extension policy in many states in India effectively discriminate against goat-owners, on the grounds that goats are particularly damaging to environment (Conway et al, 2002). This antipathy of development agencies especially within the domain of government in the name of environmental protection should change since there is no definite evidence to prove that goats pose a threat to the environment. Instead, technical, extension and marketing support should be provided so as to make goat production a profitable enterprise for all sections of society irrespective of caste, creed, land holding etc. The tradition bound communities currently not rearing goat though economically poor should be motivated through educational and incentive based developmental interventions to take up goat husbandry for their upliftment. Use of goat milk also should be promoted among the local people, and stall feeding needs encouragement. Government agencies and NGOs have to play a vital role in these efforts.

13

Technologies for Livestock Production

Growing human population, rising per capita income, and increasing urbanization are fuelling rapid growth in the demand for food of animal origin in the developing countries (Delgado et al. 1999). However, current per capita consumption is low. For instance, in India in 1993–94, per capita annual consumption of milk was 51 kg and meat 1.7 kg, much less than the world average of 75 kg milk and 34 kg meat. By 2020 the per capita consumption of milk is likely to more than double and that of meat more than triple.

On the supply side, production of both milk and meat has increased at a rate of about 5% annum. If these trends continue, increases in demand would be met adequately from domestic supplies. The sustainability of these trends, however, is uncertain. In the past, expanding livestock populations mainly contributed to the observed increases in production. India has a huge livestock population comprising different species, and further increase in the livestock population would be constrained severely by the declining land availability. Productivity of Indian livestock is low compared to many developed and developing countries. Cattle milk yield in India is about 12–15% that in the USA, Canada, and Israel. Meat yield of sheep and goats is about 60% less. Feed and fodder scarcity has been the main constraint in raising livestock productivity. Most of the feed requirement is met from crop by-products and grazing on common lands. The latter, however, have been dwindling quantitatively as well as qualitatively. The levels of adoption of breeding-, feed and nutrition-, and health-related technologies are low. It is imperative to raise these since future growth in the livestock subsect or has to come from technological changes.

The importance of livestock in India goes beyond the function of food production. It is an important source of draught power, manure for crop production and fuel for domestic use. Thus, by minimizing use of nonrenewable energy, livestock make a positive contribution to the environment. Although crops and livestock are interdependent to a large extent, the latter constitute an important mechanism for coping with the risks of crop failure. In land-scarce economies livestock provide livelihood support in terms of income and employment generation to the millions of landless and small land holders. In India, livestock wealth is mainly concentrated among the majority of marginal and small land holders.

Technology-induced growth in the livestock subsect or would thus improve food and nutritional security, alleviate poverty, and reduce inter regional and interpersonal economic inequities. India spends only about 0.5% of the agricultural gross domestic product on agricultural research. This is low compared to the average of developing countries (0.7%) and developed countries (2.5%). Livestock research receives about 20% of the total agricultural research resources. This corresponds to the contribution of livestock to the agricultural gross domestic product.

Despite the low intensity of investment in research, animal science research over the last few decades has generated a number of technologies in the areas of animal genetics and breeding, feed and nutrition, health, and management. The technical feasibility of many of these has been proven under experimental conditions. Examples include crossbreeding in cattle, sheep, pigs, and poultry; chemical and biological treatment of cereal straws; and vaccines again strinderpest, influenza, and foot and mouth disease.

Studies on returns to investment in livestock research are limited. However, sporadic evidences indicate a very high payoff to investment in livestock research and development. Despite this, the application of many technologies in the field remains limited. Except for crossbreeding, not much information is available regarding adoption and impact of other technologies. There is, thus, considerable scope to raise the productivity of livestock through application of the existing technologies.

Adoption of Technologies

Over the past few decades, animal science research has offered a number of technological options that could raise the productivity of different livestock species if adopted area-wide. These include genetic enhancement of indigenous breeds through crossbreeding with exotic breeds, improvement of nutritive quality of feed and fodder through biological and chemical treatments, development of vaccines against animal diseases, improved livestock management practices, and post harvest management. Additionally, processing technologies have been developed to strengthen the vertical linkages between the farm and dairy industry. The adoption pattern of these technologies varies widely across species, farm typologies and regions.

Breeding Technologies

Genetics and breeding research have evolved many new breeds of cattle, pig, sheep, and poultry using crossbreeding techniques. These breeds have better production coefficients compared to indigenous ones. However, their adoption in the field is limited and sporadic; only about 8% of cattle, 5% of sheep, 15% of pigs, and

33% of poultry populations belong to the crossbred/improved category. The adoption level is higher in urban areas compared to rural areas. In general wide scale adoption of crossbreeds is restricted due to their non-acclimatization to the tropical climates prevailing in most parts of the country.

Besides, their higher maintenance cost, lower disease resistance, and the poor success of artificial insemination (AI) are other barriers to adoption of crossbreeding. There are also species-specific constraints. The crossbred cow has to be replaced frequently to maintain the flow of benefits. Thus, frequent and higher acquisition costs, lack of disposal facilities (cattle slaughter is banned in most Indian states), and poor draught characteristics of male cattle are important impediments to wide-scale adoption of cross breeding technology in cattle. Nonetheless, under certain ecological and economic conditions, adoption of crossbreeding technology in cattle has been quite encouraging. The states of Kerala and Punjab, for instance, have a considerably higher proportion of crossbred cattle.

Sheep husbandry in India is practiced largely by the poor and is dependent on availability of grazing lands, which have been deteriorating in terms of both quantity and quality. In contrast to the indigenous sheep breeds that are capable of surviving even on sparse vegetation, the crossbred sheep require better nutrition. This also applies to pigs, which are often managed in scavenger systems.

In the dairy subsect or the buffalo could emerge as a promising alternative to crossbred cattle because of its adaptability to varied ecological conditions, higher milk yield and higher fat content, realizing a premium price. Unlike cattle, buffalo slaughter is not banned and animals have a disposal value. These factors have favoured a higher growth in buffalo population even without improved breeding interventions. However, in certain regions, nondescript low-yielding species have been upgraded using high-yielding breeds such as the Murrah.

The poultry subsect or has responded well to technological changes and has grown faster than the dairy and ruminant meat sector. Enhancement of genetic potential has been the most important factor in the growth of this sector. However, this has been complemented by health and nutrition technologies. The growth trends are more prominent in specialized periurban/urban poultry systems, because of higher demand for poultry meat and eggs in urban areas. Here, the entry of the private sector has boosted the adoption of technology and growth in poultry outputs. Backyard poultry production, however, continues to languish technologically.

Feed and Nutrition Technologies

A large number of India's livestock, particularly in the arid and semi-arid environments, suffer from inadequate feeding. The feed and fodder shortages, in fact, have been the main limiting factors in raising livestock productivity. Cereal crop residues comprise the main feed for livestock. However, these are deficient in crude protein and several other nutrients. Concentrate feeding is restricted to lactating, high-yielding bovines and work animals. Small ruminants derive their feed requirements mainly from grazing on common lands.

Animal nutrition and crop breeding (straw/stover) research has yielded many new technologies that could augment production and improve the nutritional quality of feeds and fodder. Research on breeding for higher yield and superior quality crop residues (such as in rice, wheat, sorghum, and millets) is in progress. Studies have indicated that a 1% increase in digestibility of sorghum/millet straw increases bovine milk yield by 5–6% (Kristjanson etal. 1998).

Apart from traditional techniques of fodder chopping and conservation, technologies such as urea treatment of fodder, strategic supplementation, urea molasses mineral blocks, and bypass protein use have the potential to alleviate feed and fodder scarcity. These technologies improve digestibility and palatability of feed, reduce feed requirements, avoid feed wastage, and contribute towards improving animal productivity. Some of these techniques, such as fodder chopping and bypass protein use, have long been in practice in many parts of the country, but are not practiced widely. The main constraint to large-scale adoption of nutrition technologies in general has been the lack of information to users.

The area under green fodder crops is also low; constituting no more than 5% of the gross cropped area. The growth in area under fodder crops has been sluggish in most parts of the country, except in the irrigated regions. This is are flection of the rising competition between food and fodder crops for limited land and other resources. Crop breeding research has evolved high-yielding varieties of a number of forage crops. However, these have not been adopted widely due to lack of awareness about new cultivars, nonavailability of irrigation water throughout the year, and problems of insect pests/diseases. Common grazing lands comprise an important source of grasses, and there exists considerable scope to raise the production of grasses/shrubs from these lands through technological and management interventions. Technological interventions such as reseeding with high-yielding grasses and watershed development, complemented with appropriate management interventions such as preventing encroachment, promotion of rotational grazing practices, and charging grazing fees have helped raise the productivity of common grazing resources and thereby improved animal performance.

Disease Control Technologies

Diseases reduce the production potential of livestock. In India many deadly diseases such as rinderpest, foot and mouth disease, hemorrhagic septicemia, and black quarter are major threats to profitable livestock production. Livestock disease control has undergone a paradigm shift in recent years. A number of biological products (vaccines) have been developed for preventive and curative disease management. The infrastructure for disease control has also expanded considerably. The main limitations to effective livestock health management are an inadequate focus on preventive measures, lack of medicines and equipment in the veterinary clinics, and ignorance among the farmers about the diseases and preventive measures. This is reflected in the frequent occurrence of many of these diseases in most parts of the country.

Processing Technologies

Post harvest technologies help producers to realize better gains from technological changes in the primary production sector. But the post harvest processing facilities are lacking in the country. Only about 20% of the total milk production is processed into value-added products. The bulk of it, however, *is processed into* ghee *and curds by the producers, and a large proportion* of this is consumed at source. Although considerable efforts have gone into developing infrastructure for milk processing in the cooperative sector, only about 5% of the total milk output is processed into table butter, cheese, milk powder, and baby foods.

Information on the proportion of other livestock products entering into the value-added chain is not available. There is a considerable demand for processed meat products, but it remains constricted due to inadequate processing facilities. The same applies to export of these products. Furthermore, not enough attention has been paid towards sanitary and phytosanitary measures. Slaughterhouses are often ill equipped and unhygienic.

Low-cost processing technologies have been developed for both cottage and large industries. The rising demand for processed milk products in urban and rural areas is expected to boost future adoption of processing technologies.

Impact of Technological Change

Technological change improves the production potential of livestock and is reflected in productivity growth. The improvements in production lead to increased welfare of the producers as well as consumers of livestock products, in India, although the intensity of adoption of different technologies is low, the technological changes as discussed above, together with improved management practices have contributed to the increased output of many livestock products.

Productivity and Production

In the last three decades milk production has increased at an annual rate of about 4.5%, and meat and egg production at about 5.5% each. The Total Factor Productivity (TFP) index, which is a joint measure of contribution of technology and technical efficiency, has grown at a rate of 1.4% a year since 1970, while the pre-1970 growth in TFP index was marginally negative. The growth in TFP is largely a result of yield-augmenting technological changes that have taken place in the dairy and poultry subsectors. The milk yield of cattle and buffalo has grown at a rate of 3.2% and 1.9%, respectively. In the case of poultry the egg yield has almost doubled and the feed conversion efficiency in broiler production has improved tremendously. The growth in productivity of species such as sheep and goats has been negligible.

Consumption, Prices, and Trade

Per capita consumption of livestock products has increased in the last three decades. The share of livestock products in food expenditure has almost doubled to 21% since 1972–73. This is due partly to increases in the availability of livestock products and a decline in prices of major products such as milk and eggs. The real prices of milk, eggs, and pork have declined, partly due to technology-driven growth in outputs. The real prices of products of sheep and goat, where technology uptake had been lacking, witnessed an upward movement.

Improvements in productivity of dairy animals have also helped achieve self-sufficiency in milk production. The imports of milk and milk products have declined to almost zero. Also, exports of certain livestock products are increasing.

Poverty and Equity

Livestock are an important source of income for the rural poor. The growth in the livestock subsect or is expected to contribute to poverty alleviation, as the livestock wealth is largely concentrated among the marginal and small land holders. These categories of farmers, however, face the problem of feed and fodder scarcity. Technologies, particularly those related to nutrition and health, are not capital intensive and could easily be adopted by them. Nevertheless, the intensity of adoption of capital-intensive technologies such as crossbreeding cattle has been observed to be higher on the landless and marginal farms; in 1991–92 there were about 10% crossbreeds in the adult female cattle herds of large land holders, while the landless and marginal land holders maintained 20% of the crossbreeds in their herds. It is thus conjectured that technological change in the livestock subsector would generate more income and employment opportunities for the resource-poor households and contribute towards alleviation of poverty and improvements in interpersonal income distribution.

Rural-Urban Disparities

Peri-urban/urban livestock systems have long existed to cater to the urban demand for foods of animal origin. Rising urbanization and a higher growth in per capita income of the urban population is causing rapid growth in demand for food of animal origin. Developments in processing and packaging technologies have also contributed considerably to this. Thus, to meet the rapidly growing demand for milk, meat, and eggs, peri-urban livestock systems have developed much faster than the rural systems. Technology adoption and production coefficients are better in peri-urban systems. Thus this type of technological dualism is likely to strengthen with further increase in the urban demand for animal foods.

Evidence and Issues

The contribution of science and technology to the growth and development of India's agricultural sector since the mid-1960s is self-evident. Developments in research and production infrastructure and encouraging government policies acted as catalysts for the technology-driven growth in agriculture, helping the country achieve self-sufficiency in food grains and many other commodities. In other areas such as horticulture, fisheries, and livestock considerable technical progress has taken place, but impacts have been slow and sporadic, and its analysis in an economic framework has largely remained undocumented.

The contribution of livestock to income and employment generation is second only to that of crops. However, its productivity is low compared to the world average. Cattle milk yield is about half the world average of 2072 kgannum-1, and the same applies to beef and pork production. The situation becomes more depressing when compared with the developed countries. Average milk yield of cattle is 13% that of the USA and 16% that of Canada. Sheep and pork yields are about 60% less. These figures suggest considerable scope for improvement of India's livestock productivity.

Nonetheless, in recent years certain outputs of livestock such as milk, meat, and eggs have been growing at an annual rate of 5% or more. The sustainability of these trends in the long run is unclear, as the current production environment has several constraints. Feed has been a limiting constraint to ruminant livestock production and this problem is likely to continue in the near future, considering the huge dimensions of livestock populations. Common grazing lands have been deteriorating quantitatively as well qualitatively (Jodha 1991). Animals are fed on crop by-products and grasses from road sides and other marginal lands. Feeding of grains and other concentrates is inadequate, and the competition for grains will intensify with increasing human and livestock populations.

Status of Technological Change

Significant research advances have been made in the areas of animal breeding, nutrition, and health. Many research products have been found to be technically and economically viable under controlled experimental conditions, but the extent of their on-farm application has been rather low. Examples of such technologies include improved animal breeds, and chemical/biological treatment of straw and fodder. A brief review of the existing, as well as potential, technologies that could influence the growth of the livestock subsector is presented below.

Breed Improvement

Crossbreeding of low-yielding indigenous breeds with high-yielding exotic breeds has been widely acknowledged as a valuable strategy to improve animal productivity. Sporadic attempts to improve the genetic potential of different livestock species (poultry, sheep, and cattle) in India were initiated during the latter half of the nineteenth century, but no significant achievements were reported. In the 1950s, systematic crossbreeding research and development programs were initiated. Crossbreeding research has focused mainly on cattle because of their dual role of milk production and use as draught animals in the crop subsector.

Between 1982 and 1992, the population of crossbred cattle increased at an annual rate of 5.6%. On the other hand, indigenous stock seemed to be approaching stabilization. The annual growth rate of indigenous cattle was 0.5% although the female population witnessed a slightly higher growth. Low milk yield and decreasing demand for draught animals are the main factors for slow growth in indigenous stock. These trends indicate a gradual substitution of indigenous cattle by crossbreds.

Despite being an important milch species, the buffalo has not received much attention in breed improvement. Development efforts have focused on upgrading low-yielding breeds through artificial insemination. The buffalo population increased at a rate of 1.9% per year between 1982 and 1992. The female population witnessed faster growth than the male population. Its adaptability to a wide range of climatic conditions, higher milk yield compared to indigenous cattle, and price premium on milk due to its higher fat content have favoured faster growth in the buffalo population. Furthermore, the disposal value of buffalo is higher; unlike cattle there are fewer restrictions on buffalo slaughter.

The population of improved poultry grew at an annual rate of about 9%, more than double that of indigenous poultry. Technological transformation of the poultry subsector seems to be market-driven as the demand for poultry meat and eggs is income-elastic and has been rising continuously. The populations of crossbred sheep and pigs also grew faster than their indigenous counterparts.

Empirical evidence from the field proves the scientific claims of better economic performance of crossbred animals. Despite this, cross breeding technology has not gained a foothold. One of the possible reasons is non acclimatization of crossbred animals to varying climatic conditions in the country, causing a number of health- and physiology-related problems. In this context, Steane (1999) mentions that 'in using exotic breeds as a strategy for improvement it seems to have been assumed that genotype-environmental interactions do not exist or the optimal economic and sustainable crossing structure would automatically be developed. Indeed in much of Asia, there appears to be a history of introducing breeds without proper evaluation and with little or no thought given to the breeding structure which will best use the available material'.

Higher initial investment and maintenance costs also limit wide spread adoption of crossbreds. The first cross animals perform very well, but the performance of animals from subsequent crosses declines significantly. Therefore crossbred animals need to be replaced frequently in order to sustain the flow of benefits. Frequent acquisition of first crossbreds, without realizing appropriate disposal value of the subsequently crossed animals, renders cross breeding technology capital intensive. In this context, Alderman (1987) observed that more than 50% of the farmers in Karnataka maintaining crossbreeds depend on the market for replacement of first cross animals. This is to avoid the risk of getting unwanted male calves and the associated problems of breeding and feeding the calf.

As indigenous cattle are sources of both milk and draught power for agriculture, their replacement by crossbred cattle has been slow. Cross bred males are considered inefficient for draught purposes compared to indigenous males, although animal physiology research has shown that the difference is marginal. Furthermore, machines have emerged as a major source of power in Indian agriculture, although mechanization does not appear to have affected the population of working animals significantly. The number of tractors per thousand ha increased from 0.6 in 1972 to 10.9 in 1992, while the working bovines declined marginally.

Mechanization has affected working bovines mainly on medium and large farms. The number of working bovines declined from 410 in 1972 to 180 per 1000 ha in 1992 on medium farms and from 210 to 50 on large farms. On small and marginal farms, most of which did not have tractors, working animals provided most of the draught power requirement in 1972. In 1992, the number of tractors on these farms had increased to 4.1 and 7.8 per thousand ha without a concomitant decline in the number of working animals.

Nonetheless, crossbreeding strategies have been successful under certain environments and economic conditions as in Kerala and Punjab. In Kerala the milk production system is cattle-based. Indigenous breeds are poor as milk producers and for draught power (Rajapurohit 1979). Cropping pattern is largely plantation-oriented, requiring less draught power. Furthermore, unlikein many other states, cattle slaughter is not prohibited by law in Kerala, thus making it easier to cull out the low-yielding and unwanted animals. In Punjab, on the other hand, increasing intensification of agriculture required more power to perform agricultural operations on time, which indigenous cattle were not capable of providing. Moreover, feed resources have never been a problem in Punjab.

Feed and Nutrition

Adequate provision of feed is essential to livestock production, and its scarcity has been one of the major limiting factors in improving productivity in India. Crop residues and by-products comprise the main feeds, accounting for 40% of the total consumption (World Bank 1996). Green fodder contributes 26%, while concentrates contribute only 3 percent. The rest comes from grazing. Stall-feeding is largely confined to buffalo, crossbred cattle, and draught animals. Small ruminants mainly feed on grasses from village common property resources, road sides, and harvested fields, and supplementary feeding is lacking except in large commercial herds.

Role of Women in Livestock Farming

As farmers, women in subsistence production ensure the survival of millions of people in all regions. Women in sustenance economies, producing and reproducing wealth in partnership with nature, are experts in their own right with ecological knowledge of nature's processes. Women's livelihood strategies, and their support and means of food security are diverse and complex, from cultivating field crops to livestock rearing, to home gardening, gathering, fishing from sources such as swamps, forests, woodlands, wastelands and quarries, etc. But these alternative modes of knowledge and livelihoods are not recognised by conventional agricultural scientists and development experts, who fail to see the connection of women's knowledge, work and skills with ensuring community food security, and the creation of wealth.

More than half of the world's food is grown by women. Women's work is both wide-ranging and multifaceted throughout the year, and they perform multiple tasks in the sphere of agriculture. Women's indigenous knowledge and skills are vitally necessary for food production and sustainable agriculture. Women's intimate knowledge of seed preparation and soil management, plants and pest control, post-harvest processing and storage, animal husbandry, as well as food processing and meal preparation are significant-crucial also to ensuring food security through sustainable agriculture. They are the authority on the interface of livestock keeping with farming.

However, there is little recognition of their significant role and contribution to the socio-economic development of a nation. The entrenched social and religious norms that define women's role as secondary and subordinate keep women vulnerable and dependent and allow women's exploitation as agricultural workers and farmers. Ensuring that the majority of rural women do not do not own land nor have access to productive resources.

As a result of rapid globalization, this patriarchal construct of women's role, position in society and the gender division of labour, in terms of production and reproduction, continue to operate. For example, women in rural agriculture face loss of livelihoods with the dumping of subsidised imported food products into her country. As a result, women who traditionally have the gendered responsibility of household food security are forced to work longer hours, and even harder to ensure the food security of the family. Women are faced with increased costs of food production as inputs costs rise, and women's health is increasingly compromised from exposure to hazardous pesticides. All these factors have intensified rural women's impoverishment, displacement and hunger. Another example, during the East Asian economic crisis, women again had to put more effort in trying to make up for lost urban incomes. In the absence of state-schemes of income and employment support, it was largely left to the so-called coping strategies of women to deal with the rural effects of the Asian crisis.

Globalization in agriculture has further intensified the development of the corporate agriculture and contract farming that perpetuates monocultures and high input agriculture. It will also intensify the control of agriculture production and distribution, and inputs into the hands of a few Transnational Corporations (TNCs) with headquarters in the North.

Women are also exploited as workers in the corporate farms that are expanding into rural areas. In these large farms or plantations, it is the women worker who is sought out and hired. Employers see women agricultural workers, as unskilled workers who will accept low wages and increased workloads without complaining and women are known to rarely join unions or organise for their rights. Women are more exposed to pesticides that adversely impact their health and their family and community's health.

This system of agriculture undoubtedly marginalize women's knowledge and skills with the introduction of new philosophies and technologies-thus eroding the base of whatever little power they had traditionally. For women it means loss of control over their knowledge and resources including seeds, plants, herbs and other devices for pest management, and commons and forest land for food and fodder gathering. Such knowledge will either become obsolete in the age of corporate agriculture or "stolen" for commercial distribution by TNCs. Productive resources will be corporatized in the name of development and globalization.

Environmental Impacts of Livestock Production

Like many industries, Industrial Farm Animal Production (ifap) results in a number of environmental impacts that affect populations both near and far. While every industry may contribute to society via production of some necessary or desired good, as our population increases, we have become more and more aware of the finite nature of our world's resources and of the impacts of our various industries upon those resources and our own human health. Industrial farm operations impact all major environmental media, including water, soil, and air. Of most concern are the pollution of ground and surface water resources with nutrients, industrial and agricultural chemicals, and microorganisms; the use of freshwater resources; the contamination and degradation of soil; and the release of toxic gases and odorous substances, as well as particulates and bioaerosols containing microorganisms and pathogens. The Commission queried the authors of this report on the magnitude and key determinants of these impacts, and the resulting impacts on both human health and ecosystems.

The major causes of the above noted environmental impacts of ifap are the enormous amounts of waste that are produced in a very small area in this agricultural model, the inadequate systems we now have to deal with that waste, and the large energy and resource inputs required for this type of production, including feed production and transport.

The usda Agricultural Research Services (ars) estimated the manure output from farm animals in the United States to be nearly 1 million US short tons of dry matter per day in 2001. Eighty-six percent of this was estimated to be produced by animals held in confinement. Different groups have posited both lower and higher estimates, but the fact remains that food animals produce an enormous amount of waste every day, exceeding human sanitary waste production by at least one order of magnitude. However, disposal of this waste is far less closely regulated than disposal of human waste. Animal manure and other agricultural waste result in water and air degradation, which in turn impact both the aquatic and the terrestrial ecosystems surrounding these operations.

In addition to the enormous waste produced by industrial agriculture, this system requires major inputs of both energy and resources. Water use is more significant in these systems because it is often used for cleaning the buildings and in the waste management systems.

In addition, the industrial model utilizes feed, which is grown in monocultures, often far away from the facility. Enormous quantities of both water and petroleum-based pesticides may be used in the production of this feed, leading not only to the depletion of water resources, but also to soil erosion and pollution with pesticides. Pesticide residues may remain in the animal feed, leading to the possibility of toxic residues in the food animals themselves. Feed crop monocultures also contribute to loss of biodiversity, as they are planted in place of other plants and/or animal habitats.

Finally, but growing more urgent every day, industrial agriculture may be a significant contributor to climate change, as the production of greenhouse gases from these facilities (both from the animals themselves and from the decomposition of their waste) is significant.

Taken together, these data suggest that the present industrial model of farm animal production is not sustainable for the long term. The overuse and degradation of natural resources may be too great to allow the current form of this production model to continue to be viable. The commission requested that the authors of this report investigate the scope of these environmental factors, to help grasp the breadth of the possible impacts of the ifap system.

Global meat production has tripled in the past three decades and could double its present level by 2050, according to a new report on the livestock industry by an international team of scientists and policy experts. The impact of this "livestock revolution" is likely to have significant consequences for human health, the environment and the global economy, the authors conclude.

Among the key findings in the report are:

- More than 1.7 billion animals are used in livestock production worldwide and occupy more than one-fourth of the Earth's land.

- Production of animal feed consumes about one-third of total arable land.
- Livestock production accounts for approximately 40 percent of the global agricultural gross domestic product.
- The livestock sector, including feed production and transport, is responsible for about 18 percent of all greenhouse gas emissions worldwide.

Impacts on Humanity

Although about 1 billion poor people worldwide derive at least some part of their livelihood from domesticated animals, the rapid growth of commercialized industrial livestock has reduced employment opportunities for many, according to the report. In developing countries, such as India and China, large-scale industrial production has displaced many small, rural producers, who are under additional pressure from health authorities to meet the food safety standards that a globalized marketplace requires.

Beef, poultry, pork and other meat products provide one-third of humanity's protein intake, but the impact on nutrition across the globe is highly variable, according to the report. "Too much animal-based protein is not good for human diets, while too little is a problem for those on a protein-starved diet, as happens in many developing countries," Mooney noted.

While over consumption of animal-source foods – particularly meat, milk and eggs – has been linked to heart disease and other chronic conditions, these foods remain a vital source of protein and nutrient nutrition throughout the developing world, the report said. The authors cited a recent study of Kenyan children that found a positive association between meat intake and physical growth, cognitive function and school performance.

Human health also is affected by pathogens and harmful substances transmitted by livestock, the authors said. Emerging diseases, such as highly pathogenic avian influenza, are closely linked to changes in the livestock production but are more difficult to trace and combat in the newly globalized marketplace, they said.

Environmental Impacts

The livestock sector is a major environmental polluter, the authors said, noting that much of the world's pasture land has been degraded by grazing or feed production, and that many forests have been clear-cut to make way for additional farmland. Feed production also requires intensive use of water, fertilizer, pesticides and fossil fuels, added co-editor Henning Steinfeld of the United Nations Food and Agriculture Organization (FAO).

A new report from FAO says livestock production is one of the major causes of the world's most pressing environmental problems, including global warming, land degradation, air and water pollution, and loss of biodiversity. Using a methodology that considers the entire commodity chain, it estimates that livestock are responsible for 18 percent of greenhouse gas emissions, a bigger share than that of transport. However, the report says, the livestock sector's potential contribution to solving environmental problems is equally large, and major improvements could be achieved at reasonable cost.

Based on the most recent data available, Livestock's long shadow takes into account the livestock sector's direct impacts, plus the environmental effects of related land use changes and production of the feed crops animals consume. It finds that expanding population and incomes worldwide, along with changing food preferences, are stimulating a rapid increase in demand for meat, milk and eggs, while globalization is boosting trade in both inputs and outputs.

In the process, the livestock sector is undergoing a complex process of technical and geographical change. Production is shifting from the countryside to urban and peri-urban areas, and towards sources of animal feed, whether feed crop areas or transport and trade hubs where feed is distributed. There is also a shift in species, with accelerating growth in production of pigs and poultry (mostly in industrial units) and a slow-down in that of cattle, sheep and goats, which are often raised extensively. Today, an estimated 80 percent of growth in the livestock sector comes from industrial production systems. Owing to those shifts, the report says, livestock are entering into direct competition for scarce land, water and other natural resources.

Deforestation, Greenhouse Gases

The livestock sector is by far the single largest anthropogenic user of land. Grazing occupies 26 percent of the Earth's terrestrial surface, while feed crop production requires about a third of all arable land. Expansion of grazing land for livestock is a key factor in deforestation, especially in Latin America: some 70 percent of previously forested land in the Amazon is used as pasture, and feed crops cover a large part of the reminder. About 70 percent of all grazing land in dry areas is considered degraded, mostly because of overgrazing, compaction and erosion attributable to livestock activity.

At the same time, the livestock sector has assumed an often unrecognized role in global warming. Using a methodology that considered the entire commodity chain, FAO estimated that livestock are responsible for 18 percent of greenhouse gas emissions, a bigger share than that of transport. It accounts for nine percent of anthropogenic carbon dioxide emissions, most of it due to expansion of pastures and arable land for feed crops. It generates even bigger shares of emissions of other gases with greater potential to warm the atmosphere: as much as 37 percent of anthropogenic methane, mostly from enteric fermentation by ruminants, and 65 percent of anthropogenic nitrous oxide, mostly from manure.

Livestock production also impacts heavily the world's water supply, accounting for more than 8 percent of global human water use, mainly for the irrigation of feed crops. Evidence suggests it is the largest sectoral source of water pollutants, principally animal wastes, antibiotics, hormones, chemicals from tanneries, fertilizers and pesticides used for feed crops, and sediments from eroded pastures.

While global figures are unavailable, it is estimated that in the USA livestock and feed crop agriculture are responsible for 37 percent of pesticide use, 50 percent of antibiotic use, and a third of the nitrogen and phosphorus loads in freshwater resources. The sector also generates almost two-thirds of anthropogenic ammonia, which contributes significantly to acid rain and acidification of ecosystems.

The sheer quantity of animals being raised for human consumption also poses a threat of the Earth's biodiversity. Livestock account for about 20 percent of the total terrestrial animal biomass, and the land area they now occupy was once habitat for wildlife.

In 306 of the 825 terrestrial eco-regions identified by the Worldwide Fund for Nature, livestock are identified as "a current threat", while 23 of Conservation International's 35 "global hotspots for biodiversity"-characterized by serious levels of habitat loss-are affected by livestock production.

In particular, water is grossly under-priced in most countries, and development of water markets and various types of cost recovery will be needed to correct the situation. In the case of land, suggested instruments include grazing fees, and better institutional arrangements for controlled and equitable access. The removal of livestock production subsidies is also likely to improve technical efficiency-in New Zealand, a drastic reduction in agricultural subsidies during the 1980s helped create one of the world's most efficient and environmentally friendly ruminant livestock industries.

Removal of price distortions at input and product level will enhance natural resource use, but may often not be sufficient. Livestock's long shadow says environmental externalities, both negative and positive, need to be explicitly factored into the policy framework. Livestock holders who provide environmental services need to be compensated, either by the immediate beneficiary (such as downstream users enjoying improved water quantity and quality) or by the general public. Services that could be rewarded include land management or land uses that restore biodiversity, and pasture management that provides for carbon sequestration. Compensation schemes also need to be developed between water and electricity providers and graziers who adopt grass lands management strategies that reduce sedimentation of water reservoirs.

Likewise, livestock holders who emit waste into waterways or release ammonia into the atmosphere should pay for the damage. Applying the "polluter pays" principle should not present insurmountable problems for offenders, given the burgeoning demand for livestock products.

Consumer Pressure

Finally, FAO says, the livestock sector is usually driven by diverse policy objectives, and decision-makers find it difficult to address economic, social, health and environmental issues at the same time. The fact that so many people depend on livestock for their livelihoods limits the policy options available, and leads to difficult and politically sensitive trade-offs.

Information, communication and education will play critical roles in enhancing a "willingness to act". With their strong and growing influence, consumers are likely to be the main source of commercial and political pressure "to push the livestock sector into more sustainable forms", Livestock's long shadow says.

Already, growing awareness of threats to the environment is translating into rising demand for environmental services: "This demand will broaden from immediate concerns-such as reducing the nuisance of flies and odours-to intermediate demands for clean air and water, then to the broader, longer-term environmental concerns, including climate change and loss of biodiversity".

Applications of Animal Biotechnology

Animal biotechnology is the application of scientific and engineering principles to the processing or production of materials by animals or aquatic species to provide goods and services (NRC 2003). Examples of animal biotechnology include generation of transgenic animals or transgenic fish (animals or fish with one or more genes introduced by human intervention), using gene knockout technology to generate animals in which a specific gene has been inactivated, production of nearly identical animals by somatic cell nuclear transfer (also referred to as clones), or production of infertile aquatic species.

Transgenics

Since the early 1980s, methods have been developed and refined to generate transgenic animals or transgenic aquatic species. For example, transgenic livestock and transgenic aquatic species have been generated with increased growth rates, enhanced lean muscle mass, enhanced resistance to disease or improved use of dietary phosphorous to lessen the environmental impacts of animal manure.

Transgenic poultry, swine, goats, and cattle also have been produced that generate large quantities of human proteins in eggs, milk, blood, or urine, with the goal of using these products as human pharmaceuticals. Examples of human pharmaceutical proteins include enzymes, clotting factors, albumin, and antibodies. The major factor limiting widespread use of transgenic animals in agricultural production systems is the relatively inefficient rate (success rate less than 10 percent) of production of transgenic animals.

Gene Knockout Technology

Animal biotechnology also can knock out or inactivate a specific gene. Knockout technology creates a possible source of replacement organs for humans. The process of transplanting cells, tissues, or organs from one species to another is referred to as "xenotransplantation." Currently, the pig is the major animal being considered as a xenotransplant donor to humans. Unfortunately, pig cells and human cells are not immunologically compatible. Pig cells express a carbohydrate epitope (alpha1, 3 galactose) on their surface that is not normally found on human cells. Humans will generate antibodies to this epitope, which will result in acute rejection of the xenograft. Genetic engineering is used to knock out or inactivate the pig gene (alpha1, 3 galactosyl transferase) that attaches this carbohydrate epitope on pig cells. Other examples of knockout technology in animals include inactivation of the prion-related peptide (PRP) gene that may generate animals resistant to diseases associated with prions (bovine spongiform encephalopathy [BSE], Creutzfeldt-Jakob Disease [CJD], scrapie, etc.). Most of the funding for these types of projects is conducted by private companies or in academic laboratories supported by the National Institutes of Health. Research projects designed to provide basic information regarding mechanisms associated with gene knockout technology are supported by NIFA.

Somatic Cell Nuclear Transfer

Another application of animal biotechnology is the use of somatic cell nuclear transfer to produce multiple copies of animals that are nearly identical copies of other animals (transgenic animals, genetically superior animals, or animals that produce high quantities of milk or have some other desirable trait, etc.). This process has been referred to as cloning. To date, somatic cell nuclear transfer has been used to clone cattle, sheep, pigs, goats, horses, mules, cats, rats, and mice. The technique involves culturing somatic cells from an appropriate tissue (fibroblasts) from the animal to be cloned. Nuclei from the cultured somatic cells are then microinjected into an enucleated oocyte obtained from another individual of the same or a closely related species. Through a process that is not yet understood, the nucleus from the somatic cell is reprogrammed to a pattern of gene expression suitable for directing normal development of the embryo. After further culture and development in vitro, the embryos are transferred to a recipient female and ultimately will result in the birth of live offspring. The success rate for propagating animals by nuclear transfer is often less than 10 percent and depends on many factors, including the species, source of the recipient ova, cell type of the donor nuclei, treatment of donor cells prior to nuclear transfer, the techniques employed for nuclear transfer, etc. NIFA has supported research projects to obtain a better understanding of the basic cellular mechanisms associated with nuclear reprogramming.

Production of Infertile Aquatic Species. In aquaculture production systems, some species are not indigenous to a given area and can pose an ecological risk to native species should the foreign species escape confinement and enter the natural ecosystem. Generation of large populations of sterile fish or mollusks is one potential solution to this problem. Techniques have been developed to alter the chromosome complement to render individual fish and mollusks infertile.

For example, triploid individuals (with three, instead of two, sets of chromosomes) have been generated by using various procedures to interfere with the final step in meiosis (extrusion of the second polar body). Timed application of high or low temperatures, various chemicals, or high hydrostatic pressure to newly fertilized eggs has been effective in producing triploid individuals. At a later time, the first cell division of the zygote can be suppressed to produce a fertile tetraploid individual (four sets of chromosomes). Tetraploids can then be mated with normal diploids to produce large numbers of infertile triploids. Unfortunately, in a commercial production system, it is often difficult to obtain sterilization of 100 percent of the individuals; thus, alternative methods are needed to ensure reproductive confinement of transgenic fish. Another technique that is being developed for finfish is to farm monosex fish stocks. Monosex populations can be produced by gender reversal and progeny testing to identify XX males for producing all female stocks or YY males for producing all male stocks.

As with any new technology, animal biotechnology faces a variety of uncertainties, safety issues and potential risks. For example, concerns have been raised regarding: the use of unnecessary genes in constructs used to generate transgenic animals, the use of vectors with the potential to be transferred or to otherwise contribute sequences to other organisms, the potential effects of genetically modified animals on the environment, the effects of the biotechnology on the welfare of the animal, and potential human health and food safety concerns for meat or animal products derived from animal biotechnology. Before animal biotechnology will be used widely by animal agriculture production systems, additional research will be needed to determine if the benefits of animal biotechnology outweigh these potential risks. The USDA Biotechnology Risk Assessment Grants program supports environmental risk assessment research projects on genetically engineered animalsAdvances in animal biotechnology have been facilitated by recent progress in sequencing and analyzing animal genomes, identification of molecular markers (microsatellites, expressed sequence tags [ESTs], quantitative trait loci [QTLs], etc.) and a better understanding of the mechanisms that regulate gene expression.

A Long History

The animal biotechnology in use today is built on a long history. Some of the first biotechnology in use includes traditional breeding techniques that date back to 5000 B.C.E. Such techniques include crossing diverse strains of animals (known as hybridizing) to produce greater genetic variety. The offspring from these crosses then are bred selectively to produce the greatest number of desirable traits. For example, female horses have been bred with male donkeys to produce mules, and male horses have been bred with female donkeys to produce hinnies, for use as work animals, for the past 3,000 years. This method continues to be used today.

The modern era of biotechnology began in 1953, when American biochemist James Watson and British biophysicist Francis Crick presented their double-helix model of DNA. That was followed by Swiss microbiologist Werner Arber's discovery in the 1960s of special enzymes, called restriction enzymes, in bacteria. These enzymes cut the DNA strands of any organism at precise points. In 1973, American geneticist Stanley Cohen and American biochemist Herbert Boyer removed a specific gene from one bacterium and inserted it into another using restriction enzymes. That event marked the beginning of recombinant DNA technology, or genetic engineering. In 1977, genes from other organisms were transferred to bacteria, an achievement that led eventually to the first transfer of a human gene.

The Technology Involved

Other Technologies

Animal biotechnology in use today is based on the science of genetic engineering. Under the umbrella of genetic engineering exist other technologies, such as transgenics and cloning, that also are used in animal biotechnology.

In addition to the use of transgenics and cloning, scientists can use gene knock out technology to inactivate, or "knock out," a specific gene. It is this technology that creates a possible source of replacement organs for humans. The process of transplanting cells, tissues or organs from one species to another is referred to as xenotransplantation. Currently, the pig is the major animal being considered as a viable organ donor to humans. Unfortunately, pig cells and human cells are not immunologically compatible. Pigs, like almost all mammals, have markers on their cells that enable the human immune system to recognize them as foreign and reject them. Genetic engineering is used to knock out the pig gene responsible for the protein that forms the marker to the pig cells.

Cloning

Scientists use reproductive cloning techniques to produce multiple copies of mammals that are nearly identical copies of other animals, including transgenic animals, genetically superior animals and animals that produce high quantities of milk or have some other desirable trait.

To date, cattle, sheep, pigs, goats, horses, mules, cats, rats and mice have been cloned, beginning with the first cloned animal, a sheep named Dolly, in 1996. Reproductive cloning begins with somatic cell nuclear transfer (SCNT).

In SCNT, scientists remove the nucleus from an egg cell (oocyte) and replace it with a nucleus from a donor adult somatic cell, which is any cell in the body except for an oocyte or sperm. For reproductive cloning, the embryo is implanted into a uterus of a surrogate female, where it can develop into a live being.

Its Applications

Animal biotechnology has many potential uses. Since the early 1980s, transgenic animals have been created with increased growth rates, enhanced lean muscle mass, enhanced resistance to disease or improved use of dietary phosphorous to lessen the environmental impacts of animal manure.

Transgenic poultry, swine, goats and cattle that generate large quantities of human proteins in eggs, milk, blood or urine also have been produced, with the goal of using these products as human pharmaceuticals. Human pharmaceutical proteins include enzymes, clotting factors, albumin and antibodies. The major factor limiting the widespread use of transgenic animals in agricultural production systems is their relatively inefficient production rate (a success rate of less than 10 percent).

A specific example of these particular applications of animal biotechnology is the transfer of the growth hormone gene of rainbow trout directly into carp eggs. The resulting transgenic carp produce both carp and rainbow trout growth hormones and grow to be one-third larger than normal carp.

Another example is the use of transgenic animals to clone large quantities of the gene responsible for a cattle growth hormone. The hormone is extracted from the bacterium, is purified and is injected into dairy cows, increasing their milk production by 10 to 15 percent. That growth hormone is called bovine somatotropin or BST.

Another major application of animal biotechnology is the use of animal organs in humans. Pigs currently are used to supply heart valves for insertion into humans, but they also are being considered as a potential solution to the severe shortage in human organs available for transplant procedures.

The Future of Animal Biotechnology

While predicting the future is inherently risky, some things can be said with certainty about the future of animal biotechnology. The government agencies involved in the regulation of animal biotechnology, mainly the Food and Drug Administration (FDA), likely will rule on pending policies and establish processes for the commercial uses of products created through the technology. In fact, as of March 2006, the FDA was expected to rule in the next few months on whether to approve meat and dairy products from cloned animals for sale to the public. If these animals and animal products are approved for human consumption, several companies reportedly are ready to sell milk, and perhaps meat, from cloned animals — most likely cattle and swine. It also is expected that technologies will continue to be developed in the field, with much hope for advances in the use of animal organs in human transplant operations.

Related Issues

The potential benefits of animal biotechnology are numerous and include enhanced nutritional content of food for human consumption; a more abundant, cheaper and varied food supply; agricultural land-use savings; a decrease in the number of animals needed for the food supply; improved health of animals and humans; development of new, low-cost disease treatments for humans; and increased understanding of human disease.

Yet despite these potential benefits, several areas of concern exist around the use of biotechnology in animals. To date, a majority of the American public is uncomfortable with genetic modifications to animals.

According to a survey conducted by the Pew Initiative on Food and Biotechnology, 58 percent of those polled said they opposed scientific research on the genetic engineering of animals. And in a Gallup poll conducted in May 2004, 64 percent of Americans polled said they thought it was morally wrong to clone animals.

Concerns surrounding the use of animal biotechnology include the unknown potential health effects to humans from food products created by transgenic or cloned animals, the potential effects on the environment and the effects on animal welfare.

Before animal biotechnology will be used widely by animal agriculture production systems, additional research will be needed to determine if the benefits of animal biotechnology outweigh these potential risks.

Food Safety

The main question posed about the safety of food produced through animal biotechnology for human consumption is, "Is it safe to eat?" But answering that question isn't simple.

Other questions must be answered first, such as, "What substances expressed as a result of the genetic modification are likely to remain in food?" Despite these questions, the

National Academies of Science (NAS) released a report titled Animal Biotechnology: Science-Based Concernsstating that the overall concern level for food safety was determined to be low. Specifically, the report listed three specific food concerns: allergens, bioactivity and the toxicity of unintended expression products.

The potential for new allergens to be expressed in the process of creating foods from genetically modified animals is a real and valid concern, because the process introduces new proteins. While food allergens are not a new issue, the difficulty comes in how to anticipate these adequately, because they only can be detected once a person is exposed and experiences a reaction.

Another food safety issue, bioactivity, asks, "Will putting a functional protein like a growth hormone in an animal affect the person who consumes food from that animal?" As of May 2006, scientists cannot say for sure if the proteins will.

Finally, concern exists about the toxicity of unintended expression products in the animal biotechnology process. While the risk is considered low, there is no data available. The NAS report stated it still needs be proven that the nutritional profile does not change in these foods and that no unintended and potentially harmful expression products appear.

Environmental Considerations

Another major concern surrounding the use of animal biotechnology is the potential for negative impact to the environment. These potential harms include the alteration of the ecologic balance regarding feed sources and predators, the introduction of transgenic animals that alter the health of existing animal populations and the disruption of reproduction patterns and their success.

To assess the risk of these environmental harms, many more questions must be answered, such as: What is the possibility the altered animal will enter the environment? Will the animal's introduction change the ecological system? Will the animal become established in the environment? and Will it interact with and affect the success of other animals in the new community? Because of the many uncertainties involved, it is challenging to make an assessment.

To illustrate a potential environmental harm, consider that if transgenic salmon with genes engineered to accelerate growth were released into the natural environment, they could compete more successfully for food and mates than wild salmon. Thus, there also is concern that genetically engineered organisms will escape and reproduce in the natural environment. It is feared existing species could be eliminated, thus upsetting the natural balance of organisms.

Legal Implications

Regulations

The regulation of animal biotechnology currently is performed under existing government agencies. To date, no new regulations or laws have been enacted to deal with animal biotechnology and related issues.

The main governing body for animal biotechnology and their products is the FDA. Specifically, these products fall under the new animal drug provisions of the Food, Drug, and Cosmetic Act (FDCA). In this use, the introduced genetic construct is considered the "drug." This lack of concrete regulatory guidance has produced many questions, especially because the process for bringing genetically engineered animals to market remains unknown.

Currently, the only genetically engineered animal on the market is the GloFish, a transgenic aquarium fish engineered to glow in the dark. It has not been subject to regulation by the FDA, however, because it is not believed to be a threat to the environment.

Many people question the use of an agency that was designed specifically for drugs to regulate live animals. The agency's strict confidentiality provisions and lack of an environmental mandate in the FDCA also are concerns. It still is unclear how the agency's provisions will be interpreted for animals and how multiple agencies will work together in the regulatory system.

When animals are genetically engineered for biomedical research purposes (as pigs are, for example, in organ transplantation studies), their care and use is carefully regulated by the Department of Agriculture. In addition, if federal funds are used to support the research, the work further is regulated by the Public Health Service Policy on

Humane Care and use of Laboratory Animals

Labeling

Whether products generated from genetically engineered animals should be labelled is yet another controversy surrounding animal biotechnology. Those opposed to mandatory labeling say it violates the government's traditional focus on regulating products, not processes. If a product of animal biotechnology has been proven scientifically by the FDA to be safe for human consumption and the environment and not materially different from similar products produced via conventional means, these individuals say it is unfair and without scientific rationale to single out that product for labeling solely because of the process by which it was made.

On the other hand, those in favour of mandatory labeling argue labeling is a consumer "right-to-know" issue. They say consumers need full information about products in the marketplace — including the processes used to make those products — not for food safety or scientific reasons, but so they can make choices in line with their personal ethics.

Intellectual Property

On average, it takes seven to nine years and an investment of about $55 million to develop, test and market a new genetically engineered product. Consequently, nearly all researchers involved in animal biotechnology are protecting their investments and intellectual property through the patent system. In 1988, the first patent was issued on a transgenic animal, a strain of laboratory mice whose cells were engineered to contain a cancer-predisposing gene. Some people, however, are opposed ethically to the patenting of life forms, because it makes organisms the property of companies. Other people are concerned about its impact on small farmers. Those opposed to using the patent system for animal biotechnology have suggested using breed registries to protect intellectual property.

Ethical and Social Considerations

Ethical and social considerations surrounding animal biotechnology are of significant importance. This especially is true because researchers and developers worry the future market success of any products derived from cloned or genetically engineered animals will depend partly on the public's acceptance of those products.

Animal biotechnology clearly has its skeptics as well as its outright opponents. Strict opponents think there is something fundamentally immoral about the processes of transgenics and cloning. They liken it to "playing God." Moreover, they often oppose animal biotechnology on the grounds that it is unnatural. Its processes, they say, go against nature and, in some cases, cross natural species boundaries.

Still others question the need to genetically engineer animals. Some wonder if it is done so companies can increase profits and agricultural production. They believe a compelling need should exist for the genetic modification of animals and that we should not use animals only for our own wants and needs. And yet others believe it is unethical to stifle technology with the potential to save human lives.

While the field of ethics presents more questions than it answers, it is clear animal biotechnology creates much discussion and debate among scientists, researchers and the American public. Two main areas of debate focus on the welfare of animals involved and the religious issues related to animal biotechnology.

Animal Welfare

Perhaps the most controversy and debate regarding animal biotechnology surrounds the animals themselves. While it has been noted that animals might, in fact, benefit from the use of animal biotechnology — through improved health, for example — the majority of discussion is about the known and unknown potential negative impacts to animal welfare through the process.

For example, calves and lambs produced through in vitro fertilization or cloning tend to have higher birth weights and longer gestation periods, which leads to difficult births that often require cesarean sections.

In addition, some of the biotechnology techniques in use today are extremely inefficient at producing fetuses that survive of the transgenic animals that do survive, many do not express the inserted gene properly, often resulting in anatomical, physiological or behavioural abnormalities. There also is a concern that proteins designed to produce a pharmaceutical product in the animal's milk might find their way to other parts of the animal's body, possibly causing adverse effects.

Animal "telos" is a concept derived from Aristotle and refers to an animal's fundamental nature. Disagreement exists as to whether it is ethical to change an animal's telos through transgenesis. For example, is it ethical to create genetically modified chickens that can tolerate living in small cages?

Those opposed to the concept say it is a clear sign we have gone too far in changing that animal. Those unopposed to changing an animal's telos, however, argue it could benefit animals by fitting them for living conditions for which they are not "naturally" suited. In this way, scientists could create animals that feel no pain.

Religious Issues

Religion plays a crucial part in the way some people view animal biotechnology. For some people, these technologies are considered blasphemous. In effect, God has created a perfect, natural order, they say, and it is sinful to try to improve that order by manipulating the basic ingredient of all life, DNA. Some religions place great importance on the "integrity" of species, and as a result, those religion's followers strongly oppose any effort to change animals through genetic modification.

Not all religious believers make these assertions, however, and different believers of the same religion might hold differing views on the subject. For example, Christians do not oppose animal biotechnology unanimously. In fact, some Christians support animal biotechnology, saying the Bible teaches humanity's dominion over nature. Some modern theologians even see biotechnology as a challenging, positive opportunity for us to work with God as "co-creators."

Transgenic animals can pose problems for some religious groups. For example, Muslims, Sikhs and Hindus are forbidden to eat certain foods. Such religious requirements raise basic questions about the identity of animals and their genetic makeup. If, for example, a small amount of genetic material from a fish is introduced into a melon (in order to allow it grow to in lower temperatures), does that melon become "fishy" in any meaningful sense?

Some would argue all organisms share common genetic material, so the melon would not contain any of the fish's identity. Others, however, believe the transferred genes are exactly what make the animal distinctive; therefore the melon would be forbidden to be eaten as well.

Thinking Critically

The following questions are intended for middle school and high school students studying animal biotechnology. They are designed to stimulate critical thinking about the many ethical issues involved in the use of animal biotechnology: Is it acceptable to create an animal that feels no pain through the use of animal biotechnology? Why or why not?

What are the benefits and to whom?

What about the concept of "telos," or an animal's true nature? Making an animal that is incapable of feeling pain goes against an animal's telos. Is this acceptable?

An animal has been used to create a promising new drug for the treatment of many types of cancer.

However, the process involved might bring pain to the animal used to make the drug. Is this acceptable considering the end result? Why or why not?

Should we make the animal incapable of pain in this case? Why or why not?

Companies have begun to apply for and receive patents to protect their often expensive investments in animal biotechnology. The average investment these companies make is about $55 million over seven to nine years. Based on this expense, should companies be able to patent a living organism? Why or why not?

How is this different from a patent issued to a company for a nonliving product, such as new software?

What alternatives can you think of for the patenting system?

14

Impact of Animal Husbandry Co-operatives

A large number of farmers in India depend on animal husbandry for their livelihood. In addition to supplying milk, meat, eggs, and hides, animals, mainly bullocks, are the major source of power for both farmers and drayers. Thus, animal husbandry plays an important role in the rural economy. The gross value of output from this sector was Rs 358 billion in FY 1989, an amount that constituted about 25 percent of the total agricultural output of Rs 1.4 trillion.

Production

In FY 1992, India had approximately 25 percent of the world's cattle, with a collective herd of 193 million head. India also had 110 million goats, 75 million water buffalo, 44 million sheep, and 10 million pigs. Milk production in FY 1990 was estimated to have reached 53.5 million tons, and egg production had reached a level of 23.3 billion eggs.

Dairy farming provided supplementary employment and an additional source of income to many small and marginal farmers. The National Dairy Development Board was established in 1965 under the auspices of Operation Flood at Anand, in Gujarat, to promote, plan, and organize dairy development through cooperatives; to provide consultations; and to set up dairy plants, which were then turned over to the cooperatives.

There were more than 63,000 Anand-style dairy cooperative societies with some 7.5 million members in the early 1990s. The milk produced and sold by these farmers brought Rs 320 million a day, or more than Rs 10 trillion a year. The increase in milk production permitted India to end imports of powdered milk and milk-related products. In addition, 30,000 tons of powdered milk were exported annually to neighbouring countries. There was about 50000 cows in India.

Operation Flood

Operation Flood, the world's largest integrated dairy development program, attempted to establish linkages between rural milk producers and urban consumers by organizing farmer-owned and-managed dairy cooperative societies. In the early 1990s, the program was in its third phase and was receiving financial assistance from the World Bank and commodity assistance from the European Economic Community. At that time, India had more than 64,000 dairy cooperative societies, with close to 7.7 million members. These cooperatives established a daily processing capacity of 15.5 million liters of whole milk and 727 tons of milk powder.

Dairy cooperatives range widely in size and function-some solely arrange for the sale of members' milk and provide few services, while others manufacture a wide range of products and may market their own branded products directly to consumers. Additionally, many offer supporting services for their members, such as providing field services, verifying weights and tests of milk, selling milk production equipment and supplies, and providing health insurance.

A dairy cooperative business is owned, operated, and controlled by the dairy farmers who benefit from its services. Members finance the cooperative and share in profits it earns in proportion to the volume of milk they market through the cooperative. Most dairy cooperatives are organized on a centralized basis--farmers are direct members. Only a few dairy cooperatives are organized on a federated basis—members of the cooperative are other cooperatives or a combination of direct members and cooperative members. Many are organized to serve farmers in a local area or single State, while others serve members in multiple States, regionally or nationally. Some dairy cooperatives have made additional business arrangements to increase outlets for members' milk through subsidiaries, partnerships, federations, marketing agenciesin-common with other cooperatives, and joint ventures with other cooperatives or investor-owned firms.

Dairy cooperatives represented 13 percent of all agricultural marketing cooperatives in the United States in 2002, markedly down from 30 percent in the mid-1940s. This decline has been faster than for other agricultural commodities. Likewise, the number of milk producers belonging to a dairy marketing cooperative shrunk to 61,390 in 2002 from a peak in the 1950s of around 777,000. However, this trend mirrors the national decline in total number of dairy farms and increase in dairy farm size.

In sharp contrast, the volume of milk handled by cooperatives was 144 billion pounds in 2002, having expanded steadily over the years from 31 billion pounds in the mid-1930s. On a percooperative basis, cooperatives handled an average of 736 million pounds of milk each in 2002, compared to 14 million pounds in the mid-1930s. Thus, despite their declining numbers, dairy cooperatives' role in the marketplace has continued to grow. The share of all milk delivered to plants and dealers in the

United States by cooperatives was 86 percent in 2002, up from 48 percent in the mid-1930s. Almost all of the milk received by dairy cooperatives (96 percent in 2002) came directly from member-producers. The rest came from nonmembers or investor-owned firms.

The share of milk represented by cooperatives varies between the regions of the United States. More than 90 percent of the milk sold to plants and dealers in the East North Central, West North Central, and South Atlantic regions was handled by cooperatives in 2002. In contrast, cooperatives had the lowest share in the North Atlantic region (72 percent) followed by the Western region (75 percent). (Ironically, 44 percent of all cooperatives are headquartered in the North Atlantic region.) The cooperative share was 87 percent in the South Central region.

In 2002, the majority (56 percent) of U.S. dairy cooperatives were categorized as small cooperatives, handling less than 50 million pounds of milk annually. Thirty-one percent of the Nation's dairy cooperatives were medium-sized—meaning they handled 50 to 99 million pounds of milk annually, while large cooperatives, those handling at least 1 billion pounds of milk per year, accounted for 13 percent.

Cooperatives sold 62 percent of the milk they marketed raw and processed or manufactured 38 percent in their own plants in 2002. Dairy cooperatives have marketed a majority of the cheddar cheese, butter, and dry milk products produced in the United States for about the past 60 years.

Cooperatives' share of dry milk products peaked in 1987 at 91 percent and in 2002 held an 85-percent share. Cooperative share of U.S. butter production has fluctuated between 61 and 71 percent since the 1960s and accounted for 71 percent in 2002. Shares of all natural cheese production have fallen to 40 percent in 2002 from a high of 47 percent in 1980. Minor shares of the Nation's packaged fluid milk, cottage cheese, and ice cream were distributed by cooperatives over the years.

The estimated net business volume of the Nation's dairy cooperatives has expanded 40-fold, from $520 million in the mid-1930s to $23 billion in 2002.That volume represents one-third of the total net business volume of all agricultural marketing cooperatives in the United States in 2002 and has fluctuated from 27 percent (1943-44) to 38 percent (1987).

Cooperatives in the Dairy Industry

Milk is unique among farm commodities. It is highly perishable, produced, and "harvested" on a daily basis, and moved from farm to market every other day, if not every day. The volume of milk produced varies seasonally and daily for biological reasons. This variation is not coordinated with changes in demand, which also vary from day to day and from season to season. The task of balancing, or coordinating, the amount of milk supplied with the volume of milk demanded is thus problematical.

Storage to balance supplies with demand is feasible only after processing, except in the very short term. As technology developed, conversion of milk from raw product to various intermediate and final products with longer shelf-lives became possible, but required increasingly capital-intensive facilities and technologies that are subject to significant economies of scale.

These fundamental characteristics of milk production, in concert with adverse marketing conditions and the economies available from jointly owned milk handling facilities and manufacturing plants, led dairy farmers to pioneer the application of cooperative principles to marketing U.S. farm products.

Initial Organization

In the early days of the Nation, dairy farms were relatively small and remotely located. Cooperatives sprang up spontaneously, formed by groups of farmers seeking solutions to common problems. These groups drew upon cooperative traditions that immigrant dairy farmers had brought with them from Northern Europe. Milk from several farms was pooled in one location (either by hauling milk or cream in cans or by taking cows to the factory to be milked) and made into cheese or butter.

Part of the net proceeds was returned to patrons in proportion to the amount of milk each furnished. Cooperative creameries were generally organized in areas where a large portion of the milk produced could best be marketed for butter production, thereby avoiding the high cost of transporting whole milk to distant city markets.

The first reported cooperative cheese factories were established in the mid-1800s. The number of creameries grew slowly until mechanical cream separators were introduced around 1890.

By 1900, there were around 6,000 creameries and almost 3,000 cheese factories. About one-third were organized as cooperatives.

Milk evaporating and drying facilities emerged in the 1920s and subsequently some creameries installed milk drying facilities to provide a market for buttermilk and skim milk.

Concurrently, the organized marketing of raw milk for fluid consumption began during the latter part of the 18th century in cities where families were unable to obtain milk from nearby producers.

A system of “middle-men” between producers and consumers began to emerge in the 1800s. Fewer and fewer producers carried out all marketing functions. Milk price was determined by negotiation; both buyers and sellers were small and numerous.

During the mid-1800s, the rapid construction of railroads permitted increased movement of “fresh country” milk to the cities.

Expanding urbanization made it necessary for families to obtain milk from distant dairy farms in the country. Dairy farmers formed associations to arrange these early shipments of "pure" country milk to the cities.

By the late 1800s, the milk marketing system was steadily moving toward a structure where hundreds or thousands of dairy farmers sold to only a handful of large fluid milk dealers. Consequently, cooperative associations developed around the major cities in the eastern part of the United States and in Chicago to negotiate milk prices with milk dealers and distributors. One tactic the early cooperatives employed to compel reluctant milk dealers to negotiate with them was the "milk strike."

Farmers would withhold milk from the market which would tighten supplies. This had short-term success in enforcing cooperative demands. Even so, the dealers began to develop a bargaining edge over farmers, primarily due to better market information through their powerful organizations. In addition, the rural isolation and the generally independent nature of most dairy farmers combined to restrain cooperative growth at that time.

Nonetheless, early cooperative associations laid the foundation upon which later ones were built. In the early 20th century, unfavourable economic conditions, chaotic pricing of fluid milk, and dealers who balanced fluctuating supply needs by refusing to accept some producers' milk spurred the successful formation of large-scale cooperative bargaining organizations for raw whole milk. Another important stimulus to cooperative development was government policy for food control during World War I.

Dairy Cooperative Operations

Dairy cooperatives today run the gamut in size and extent of services they provide. They differ markedly because they have taken a variety of avenues to address the needs and preferences of their members and specific market situation.

Most dairy cooperatives employ a general manager or CEO and a staff of field representatives to build membership and provide assistance in improving milk production and quality. Other employees include: laboratory technicians who verify the weight and quality of farm milk delivered; personnel to prepare and distribute market information and represent the association in legislative and regulatory affairs (such as Federal and State milk market order hearings); and office staff who prepare and distribute milk checks and perform other administrative duties.

Cooperatives that manufacture or process milk also have employees who operate the cooperative's plants and develop and market its products. Some cooperatives also provide other services such as selling dairy equipment, supplies, and feed (even to the extent of having their own system of feed mills, in some cases), providing health insurance, retirement plans, and, beginning in the 1990s, forward milk price contracts and assistance in using futures markets to manage milk price risks. Alternatively, many small cooperatives, predominately in the Northeast, limit their activities to arranging milk sales and issuing milk checks. These are often affiliated with larger cooperatives and usually have no hired staff.

A board of directors oversees every dairy cooperative. The board consists of elected producer-members who set the cooperative's policy for the manager to carry out.

Individual cooperatives have tailored their operations to their specific market situation to assure members a market for their milk. Their divergent operations can be broadly grouped into two operating types: "bargaining only" and "manufacturing/ processing."

Furthermore, within the manufacturing/processing group, there is a wide range in the type and scale of manufacturing operations.

Bargaining-only Cooperatives

Many cooperatives focus their operations on negotiating milk prices and terms of trade for members' raw milk, but do not engage in further manufacturing or processing. These "bargaining-only" dairy cooperatives are the most numerous (74 percent of all dairy cooperatives in 2002), but represent just one-quarter of U.S. cooperative milk volume. Some of the larger bargaining only cooperatives once had manufacturing capabilities, but ceased their plant operations to focus only on milk marketing. A number of them have joined in pricing federations. A small number are "check-off" organizations that represent member concerns in the policy arena and provide limited marketing services.

Cooperative

A cooperative (also co-operative; often referred to as a co-op or coop) is a business organization owned and operated by a group of individuals for their mutual benefit. Cooperatives are defined by the International Co-operative Alliance's Statement on the Co-operative Identity as autonomous associations of persons united voluntarily to meet their common economic, social, and cultural needs and aspirations through jointly-owned and democratically-controlled enterprises. A cooperative may also be defined as a business owned and controlled equally by the people who use its services or by the people who work there. Cooperative enterprises are the focus of study in the field of cooperative economics.

Origins

Cooperation dates back as far as human beings have been organising for mutual benefit. Tribes were organised as cooperative structures, allocating jobs and resources among each other, only trading with the external communities. Post-industrial Europe is home to the first co-operatives from an industrial context.

In 1761, the Fenwick Weavers' Society was formed in Fenwick, East Ayrshire, Scotland to sell discounted oatmeal to local workers. Its services expanded to include assistance with savings and loans, emigration and education. In 1810, Welsh social reformer Robert Owen, from Newtown in mid-Wales, and his partners purchased New Lanark mill from Owen's father-in-law and proceeded to introduce better labour standards including discounted retail shops where profits were passed on to his employees. Owen left New Lanark to pursue other forms of co-operative organization and develop co-op ideas through writing and lecture. Co-operative communities were set up in Glasgow, Indiana and Hampshire, although ultimately unsuccessful. In 1828, William King set up a newspaper, The Cooperator, to promote Owen's thinking, having already set up a co-operative store in Brighton.

The Rochdale Society of Equitable Pioneers, founded in 1844, is usually considered the first successful co-operative enterprise, used as a model for modern co-ops, following the 'Rochdale Principles'. A group of 28 weavers and other artisans in Rochdale, England set up the society to open their own store selling food items they could not otherwise afford. Within ten years there were over 1,000 co-operative societies in the United Kingdom.

Other events such as the founding of a friendly society by the Tolpuddle Martyrs in 1832 were key occasions in the creation of organized labour and consumer movements.

Social Economy

In the final year of the 20th century, cooperatives banded together to establish a number of social enterprise agencies which have moved to adopt the multi-stakeholder cooperative model. In the last 15 years (1994-2009) the EU and its member nations, have gradually revised national accounting systems to "make visible" the increasing contribution of social economy organizations.

Organizational and Ideological Roots

The roots of the cooperative movement can be traced to multiple influences and extend worldwide. In the Anglosphere, post-feudal forms of cooperation between workers and owners, that are expressed today as "profit-sharing" and "surplus sharing" arrangements, existed as far back as 1795.

The key ideological influence on the Anglosphere branch of the cooperative movement, however, was a rejection of the charity principles that underpinned welfare reforms when the British government radically revised its Poor Laws in 1834. As both state and church institutions began to routinely distinguish between the 'deserving' and 'undeserving' poor, a movement of friendly societies grew throughout the British Empire based on the principle of mutuality, committed to self-help in the welfare of working people.

Friendly Societies established forums through which one member, one vote was practiced in organisation decision-making. The principles challenged the idea that a person should be an owner of property before being granted a political voice. Throughout the second half of the nineteenth century (and then repeatedly every 20 years or so) there has been a surge in the number of cooperative organisations, both in commercial practice and civil society, operating to advance democracy and universal suffrage as a political principle.

Friendly Societies and consumer cooperatives became the dominant form of organization amongst working people in Anglosphere industrial societies prior to the rise of trade unions and industrial factories. Weinbren reports that by the end of the 19th century, over 80% of British working age men and 90% of Australian working age men were members of one or more Friendly Society.

From the mid-nineteenth century, mutual organisations embraced these ideas in economic enterprises, firstly amongst tradespeople, and later in cooperative stores, educational institutes, financial institutions and industrial enterprises. The common thread (enacted in different ways, and subject to the constraints of various systems of national law) is the principle that an enterprise or association should be owned and controlled by the people it serves, and share any surpluses on the basis of each members' cooperative contribution (as a producer, labourer or consumer) rather than their capacity to invest financial capital.

The cooperative movement has been fueled globally by ideas of economic democracy. Economic democracy is a socioeconomic philosophy that suggests an expansion of decision-making power from a small minority of corporate shareholders to a larger majority of public stakeholders.

There are many different approaches to thinking about and building economic democracy. Both Marxism and anarchism, for example, have been influenced by utopian socialism, which was based on voluntary cooperation, without recognition of class conflict.

Anarchists are committed to libertarian socialism and they have focused on local organization, including locally-managed cooperatives, linked through confederations of unions, cooperatives and communities.

Marxists, who as socialists have likewise held and worked for the goal of democratizing productive and reproductive relationships, often placed a greater strategic emphasis on confronting the larger scales of human organization. As they viewed the capitalist class to be prohibitively politically, militarily and culturally mobilized in order to maintain an exploitable working class, they fought in the early twentieth century to appropriate from the capitalist class the society's collective political capacity in the form of the state, either through democratic socialism, or through what came to be known as Leninism.

Though they regard the state as an unnecessarily oppressive institution, Marxists considered appropriating national and international-scale capitalist institutions and resources (such as the state) to be an important first pillar in creating conditions favourable to solidaristic economies. With the declining influence of the USSR after the 1960s, socialist strategies pluralized, though economic democratizers have not as yet established a fundamental challenge to the hegemony of global neoliberal capitalism.

Meaning

Cooperatives as legal entities:

A cooperative is a legal entity owned and democratically controlled by its members. Members often have a close association with the enterprise as producers or consumers of its products or services, or as its employees.

In some countries, e.g. Finland and Sweden, there are specific forms of incorporation for co-operatives. Cooperatives may take the form of companies limited by shares or by guarantee, partnerships or unincorporated associations. In the USA, cooperatives are often organized as non-capital stock corporations under state-specific cooperative laws. However, they may also be unincorporated associations or business corporations such as limited liability companies or partnerships; such forms are useful when the members want to allow:

1. some members to have a greater share of the control, or
2. some investors to have a return on their capital that exceeds fixed interest, neither of which may be allowed under local laws for cooperatives. Cooperatives often share their earnings with the membership as dividends, which are divided among the members according to their participation in the enterprise, such as patronage, instead of according to the value of their capital shareholdings (as is done by a joint stock company).

Identity

Cooperatives are based on the cooperative values of "self-help, self-responsibility, democracy and equality, equity and solidarity" and the seven cooperative principles.

1. Voluntary and Open Membership
2. Democratic Member Control
3. Member Economic Participation
4. Autonomy and Independence
5. Education, Training and Information
6. Cooperation among Cooperatives
7. Concern for Community.

In the tradition of their founders, cooperative members believe in the ethical values of honesty, openness, social responsibility and caring for others. Such legal entities have a range of unique social characteristics. Membership is open, meaning that anyone who satisfies certain non-discriminatory conditions may join. Economic benefits are distributed proportionally according to each member's level of participation in the cooperative, for instance by a dividend on sales or purchases, rather than divided according to capital invested.

Cooperatives may be generally classified as either consumer cooperatives or producer cooperatives. Cooperatives are closely related to collectives, which differ only in that profit-making or economic stability is placed secondary to adherence to social-justice principles. Co-ops can sometimes be identified on the Internet through the use of the.coop gTLD. Organizations using.coop domain names must adhere to the basic co-op values.

Livestock and Poultry

A large number of farmers depend on livestock for their livelihood. In addition to supplying milk, meat, eggs, and hides, animals, mainly bullocks, are the major source of power for both farmers and drayers. Thus, animal husbandry plays an important role in the rural economy. The gross value of output from this sector was Rs 358 billion in FY 1989, an amount that constituted about 25 percent of the total agricultural output of Rs 1.4 trillion.

Amul

Amul ("priceless" in Sanskrit. The brand name "Amul," from the Sanskrit "Amoolya," was suggested by a quality control expert in Anand.), formed in 1946, is a dairy cooperative in India. It is a brand name managed by an apex cooperative organisation, Gujarat Co-operative Milk Marketing Federation Ltd. (GCMMF), which today is jointly owned by some 2.8 million milk producers in Gujarat, India.

AMUL is based in Anand, Gujarat and has been an example of a co-operative organization's success in the long term. It is one of the best examples of co-operative achievement in the developing economy. "Anyone who has seen... the dairy cooperatives in the state of Gujarat, especially the highly successful one known as AMUL, will naturally wonder what combination of influences and incentives is needed to multiply such a model a thousand times over in developing regions everywhere."

The Amul Pattern has established itself as a uniquely appropriate model for rural development. Amul has spurred the White Revolution of India, which has made India the largest producer of milk and milk products in the world. It is also the world's biggest vegetarian cheese brand.

Amul is the largest food brand in India and world's Largest Pouched Milk Brand with an annual turnover of US $1050 million (2006–07). Currently Unions making up GCMMF have 2.8 million producer members with milk collection average of 10.16 million litres per day. Besides India, Amul has entered overseas markets such as Mauritius, UAE, USA, Bangladesh, Australia, China, Singapore, Hong Kong and a few South African countries. Its bid to enter Japanese market in 1994 did not succeeded, but now it has fresh plans entering the Japanese markets. Other potential markets being considered include Sri Lanka.

Dr Verghese Kurien, former chairman of the GCMMF, is recognised as a key person behind the success of Amul. On 10 Aug 2006 Parthi Bhatol, chairman of the Banaskantha Union, was elected chairman of GCMMF.

GCMMF Today

GCMMF is India's largest food products marketing organisation.. It is a state level apex body of milk cooperatives in Gujarat, which aims to provide remunerative returns to the farmers and also serve the interest of consumers by providing affordable quality products. GCMMF markets and manages the Amul brand. From mid-1990s Amul has entered areas not related directly to its core business. Its entry into ice cream was regarded as successful due to the large market share it was able to capture within a short period of time-primarily due to the price differential and the brand name. It also entered the pizza business, where the base and the recipes were made available to restaurant owners who could price it as low as 30 rupees per pizza when the other players were charging upwards of 100 rupees.

Company info

The Gujarat Cooperative Milk Marketing Federation Ltd, Anand (GCMMF) is the largest food products marketing organisation of India. It is the apex organization of the Dairy Cooperatives of Gujarat. This State has been a pioneer in organizing dairy cooperatives and our success has not only been emulated in India but serves as a model for rest of the World.

Over the last five and a half decades, Dairy Cooperatives in Gujarat have created an economic network that links more than 2.8 million village milk producers with millions of consumers in India and abroad through a cooperative system that includes 13,141 Village Dairy Cooperative Societies (VDCS) at the village level, affiliated to 13 District Cooperative Milk Producers' Unions at the District level and GCMMF at the State level. These cooperatives collect on an average 7.5 million litres of milk per day from their producer members, more than 70% of whom are small, marginal farmers and landless labourers and include a sizeable population of tribal folk and people belonging to the scheduled castes.

The turnover of GCMMF (AMUL) during 2008-09 was Rs. 67.11 billion. It markets the products, produced by the district milk unions in 30 dairy plants, under the renowned AMUL brand name. The combined processing capacity of these plants is 11.6 million litres per day, with four dairy plants having processing capacity in excess of 1 million Litres per day.

The farmers of Gujarat own the largest state of the art dairy plant in Asia – Mother Dairy, Gandhinagar, Gujarat-which can handle 2.5 million litres of milk per day and process 100 MTs of milk powder daily.

During the last year, 3.1 billion litres of milk was collected by Member Unions of GCMMF. Huge capacities for milk drying, product manufacture and cattle feed manufacture have been installed. All its products are manufactured under the most hygienic conditions. All dairy plants of the unions are ISO 9001-2000, ISO 22000 and HACCP certified. GCMMF (AMUL)'s Total Quality Management ensures the quality of products right from the starting point (milk producer) through the value chain until it reaches the consumer.

Ever since the movement was launched fifty-five years ago, Gujarat's Dairy Cooperatives have brought about a significant social and economic change to our rural people. The Dairy Cooperatives have helped in ending the exploitation of farmers and demonstrated that when our rural producers benefit, the community and nation benefits as well.

The Gujarat Cooperative Milk Marketing Federation Ltd. cannot be viewed simply as a business enterprise. It is an institution created by the milk producers themselves to primarily safeguard their interest economically, socially as well as democratically. Business houses create profit in order to distribute it to the shareholders. In the case of GCMMF the surplus is ploughed back to farmers through the District Unions as well as the village societies. This circulation of capital with value addition within the structure not only benefits the final beneficiary – the farmer – but eventually contributes to the development of the village community. This is the most significant contribution the Amul Model cooperatives has made in building the Nation.

The Birth of Amul and Development of India's Dairy Cooperative Movement

The birth of Amul at Anand provided the impetus to the cooperative dairy movement in the country. The Kaira District Cooperative Milk Producers' Union Limited was registered on December 14, 1946 as a response to exploitation of marginal milk producers by traders or agents of existing dairies in the small town named Anand (in Kaira District of Gujarat).

Milk Producers had to travel long distances to deliver milk to the only dairy, the Polson Dairy in Anand. Often milk went sour as producers had to physically carry the milk in individual containers, especially in the summer season. These agents arbitrarily decided the prices depending on the production and the season. Milk is a commodity that has to be collected twice a day from each cow/buffalo. In winter, the producer was either left with surplus/unsold milk or had to sell it at very low prices. Moreover, the government at that time had given monopoly rights to Polson Dairy (around that time Polson was the most well known butter brand in the country) to collect milk from Anand and supply it to Bombay city in turn. India ranked nowhere amongst milk producing countries in the world in 1946.

Angered by the unfair and manipulative trade practices, the farmers of Kaira District approached Sardar Vallabhbhai Patel (who later became the first Deputy Prime Minister and Home Minister of free India) under the leadership of the local farmer leader Tribhuvandas Patel.

Sardar Patel advised the farmers to form a Cooperative and supply milk directly to the Bombay Milk Scheme instead of selling it to Polson (who did the same but gave low prices to the producers). He sent Morarji Desai (who later became Prime Minister of India) to organize the farmers. In 1946, the farmers of the area went on a milk strike refusing to be further oppressed. Thus the Kaira District Cooperative was established to collect and process milk in the District of Kaira in 1946. Milk collection was also decentralized, as most producers were marginal farmers who were in a position to deliver 1-2 litres of milk per day. Village level cooperatives were established to organize the marginal milk producers in each of these villages.

The Cooperative was further developed & managed by Dr. V Kurien along with Shri H M Dalaya. The first modern dairy of the Kaira Union was established at Anand (which popularly came to be known as AMUL dairy after its brand name). Indigenous R&D and technology development at the Cooperative had led to the successful production of skimmed milk powder from buffalo milk – the first time on a commercial scale anywhere in the world. The foundations of a modern dairy industry in India were thus laid since India had one of the largest buffalo populations in the world.

The success of the dairy co-operative movement spread rapidly in Gujarat. Within a short span five other district unions – Mehsana, Banaskantha, Baroda, Sabarkantha and Surat were organized. In order to combine forces and expand the market while saving on advertising and avoid a situation where milk cooperatives would compete against each other it was decided to set up an apex marketing body of dairy cooperative unions in Gujarat.

Thus, in 1973, the Gujarat Co-operative Milk Marketing Federation was established. The Kaira District Co-operative Milk Producers' Union Ltd. which had established the brand name AMUL in 1955 decided to hand over the brand name to GCMMF (AMUL). With the creation of GCMMF (AMUL), we managed to eliminate competition between Gujarat's cooperatives while competing with the private sector as a combined stronger force. GCMMF (AMUL) has ensured remunerative returns to the farmers while providing consumers with products under the brand name AMUL. This was possible due to the leadership of the founder Chairman of AMUL, Tribhuvandas Patel and the vision of the father of the White Revolution, Dr. Verghese Kurien who worked as a professional manager at AMUL. Numerous people contributed to this movement which would otherwise not have been possible.

Dr. Verghese Kurien, the World Food Prize and the Magsaysay Award winner, is the architect of India's White Revolution, which helped India emerge as the largest milk producer in the world.

Impressed with the development of dairy cooperatives in Kaira District & its success, Shri Lal Bahadur Shastri, the then Prime Minister of India during his visit to Anand in 1964, asked Dr. V Kurien to replicate the Anand type dairy cooperatives all over India. Thus, the National Dairy Developed Board was formed and Operation Flood Programme was launched for replication of the Amul Model all over India.

Operation Flood, the world's largest dairy development programme, is based on the experience gained from the 'Amul Model' dairy cooperatives. The facilities at all levels are entirely farmer-owned.

The cooperatives are able to build markets, supply inputs and create value-added processing. Thus, Amul Model cooperatives seem to be the most appropriate organizational force for promoting agricultural development using modern technologies and professional management and thereby generating employment for the rural masses and eradicating poverty in these undeveloped areas. India has already demonstrated the superiority of this approach.

The Three-tier "Amul Model"

The Amul Model is a three-tier cooperative structure. This structure consists of a Dairy Cooperative Society at the village level affiliated to a Milk Union at the District level which in turn is further federated into a Milk Federation at the State level.

The above three-tier structure was set-up in order to delegate the various functions, milk collection is done at the Village Dairy Society, Milk Procurement & Processing at the District Milk Union and Milk & Milk Products Marketing at the State Milk Federation. This helps in eliminating not only internal competition but also ensuring that economies of scale is achieved. As the above structure was first evolved at Amul in Gujarat and thereafter replicated all over the country under the Operation Flood Programme, it is known as the 'Amul Model' or 'Anand Pattern' of Dairy Cooperatives.

Responsible for Marketing of Milk & Milk Products Responsible for Procurement & Processing of Milk Responsible for Collection of Milk Responsible for Milk Production

Village Dairy Cooperative Society (VDCS)

The milk producers of a village, having surplus milk after own consumption, come together and form a Village Dairy Cooperative Society (VDCS). The Village Dairy Cooperative is the primary society under the three-tier structure. It has membership of milk producers of the village and is governed by an elected Management Committee consisting of 9 to 12 elected representatives of the milk producers based on the principle of one member, one vote. The village society further appoints a Secretary (a paid employee and member secretary of the Management Committee) for management of the day-to-day functions. It also employs various people for assisting the Secretary in accomplishing his/her daily duties. The main functions of the VDCS are as follows:

- Collection of surplus milk from the milk producers of the village & payment based on quality & quantity
- Providing support services to the members like Veterinary First Aid, Artificial Insemination services, cattle-feed sales, mineral mixture sales, fodder & fodder seed sales, conducting training on Animal Husbandry & Dairying, etc.
- Selling liquid milk for local consumers of the village
- Supplying milk to the District Milk Union.

Thus, the VDCS in an independent entity managed locally by the milk producers and assisted by the District Milk Union.

District Cooperative Milk Producers' Union (Milk Union)

The Village Societies of a District (ranging from 75 to 1653 per Milk Union in Gujarat) having surplus milk after local sales come together and form a District Milk Union.

The Milk Union is the second tier under the three-tier structure. It has membership of Village Dairy Societies of the District and is governed by a Board of Directors consisting of 9 to 18 elected representatives of the Village Societies. The Milk Union further appoints a professional Managing Director (paid employee and member secretary of the Board) for management of the day-to-day functions. It also employs various people for assisting the Managing Director in accomplishing his/her daily duties.

The main functions of the Milk Union are as follows:

- Procurement of milk from the Village Dairy Societies of the District
- Arranging transportation of raw milk from the VDCS to the Milk Union.
- Providing input services to the producers like Veterinary Care, Artificial Insemination services, cattle-feed sales, mineral mixture sales, fodder & fodder seed sales, etc.

- Conducting training on Cooperative Development, Animal Husbandry & Dairying for milk producers and conducting specialised skill development & Leadership Development training for VDCS staff & Management Committee members.
- Providing management support to the VDCS along with regular supervision of its activities.
- Establish Chilling Centres & Dairy Plants for processing the milk received from the villages.
- Selling liquid milk & milk products within the District
- Process milk into various milk & milk products as per the requirement of State Marketing Federation.
- Decide on the prices of milk to be paid to milk producers as well on the prices of support services provided to members.

State Cooperative Milk Federation (Federation)

The Milk Unions of a State are federated into a State Cooperative Milk Federation. The Federation is the apex tier under the three-tier structure. It has membership of all the cooperative Milk Unions of the State and is governed by a Board of Directors consisting of one elected representative of each Milk Union.

The State Federation further appoints a Managing Director (paid employee and member secretary of the Board) for management of the day-to-day functions. It also employs various people for assisting the Managing Director in accomplishing his daily duties. The main functions of the Federation are as follows:

- Marketing of milk & milk products processed/manufactured by Milk Unions.
- Establish distribution network for marketing of milk & milk products.
- Arranging transportation of milk & milk products from the Milk Unions to the market.
- Creating & maintaining a brand for marketing of milk & milk products (brand building).
- Providing support services to the Milk Unions & members like Technical Inputs, management support & advisory services.
- Pooling surplus milk from the Milk Unions and supplying it to deficit Milk Unions.
- Establish feeder-balancing Dairy Plants for processing the surplus milk of the Milk Unions.
- Arranging for common purchase of raw materials used in manufacture/ packaging of milk products.

- Decide on the prices of milk & milk products to be paid to Milk Unions.
- Decide on the products to be manufactured at various Milk Unions (product-mix) and capacity required for the same.
- Conduct long-term Milk Production, Procurement & Processing as well as Marketing Planning.
- Arranging Finance for the Milk Unions and providing them technical know-how.
- Designing & Providing training on Cooperative Development, Technical & Marketing functions.
- Conflict Resolution & keeping the entire structure intact.

We move to the year 2008. The dairy industry in India and particularly in the State of Gujarat looks very different. India for one has emerged as the largest milk producing country in the World. Gujarat has emerged as the most successful State in terms of milk and milk product production through its cooperative dairy movement. The Kaira District Cooperative Milk Producers' Union Limited, Anand has become the focal point of dairy development in the entire region and AMUL has emerged as one of the most recognized brands in India, ahead of many international brands.

Today, we have around 176 cooperative dairy Unions formed by 1,25,000 dairy cooperative societies having a total membership of around 13 million farmers on the same pattern, who are processing and marketing milk and milk products profitably, be it Amul in Gujarat or Verka in Punjab, Vijaya in Andhra Pradesh or a Nandini in Karnataka. This entire process has created more than 190 dairy processing plants spread all over India with large investments by these farmers' institutions. These cooperatives today collect approximately 23 million kgs. of milk per day and pay an aggregate amount of more than Rs.125 billion to the milk producers in a year.

Impact of the "Amul Model"

The effects of Operation Flood Programme are more appraised by the World Bank in its recent evaluation report. It has been proved that an investment of Rs. 20 billion over 20 years under Operation Flood Programme in 70s & 80s has contributed in increase of India's milk production by 40 Million Metric Tonne (MMT) i.e. from about 20 MMT in pre-Operation Flood period to more than 60 MMT at the end of Operation flood Programme. Thus, an incremental return of Rs. 400 billion annually have been generated by an investment of Rs. 20 billion over a period of 20 years. This has been the most beneficial project funded by the World Bank anywhere in the World. One can continue to see the effect of these efforts as India's milk production continues to increase and now stands at 90 MMT. Despite this fourfold increase in milk production, there has not been drop in the prices of milk during the period and has continued to grow.

Due to this movement, the country's milk production tripled between the years 1971 to 1996. Similarly, the per capita milk consumption doubled from 111 gms per day in 1973 to 222 gms per day in 2000.

Thus, these cooperatives have not just been instrumental in economic development of the rural society of India but it also has provided vital ingredient for improving health & nutritional requirement of the Indian society. Very few industries of India have such parallels of development encompassing such a large population.

These dairy cooperatives have been responsible in uplifting the social & economic status of the women folk in particular as women are basically involved in dairying while the men are busy with their agriculture. This has also provided a definite source of income to the women leading to their economic emancipation.

The three-tier 'Amul Model' has been instrumental in bringing about the White Revolution in the country. As per the assessment report of the World Bank on the Impact of Dairy Development in India, the 'Anand Pattern' has demonstrated the following benefits:

- The role of dairying in poverty reduction
- The fact that rural development involves more than agricultural production
- The value of national 'ownership' in development
- The beneficial effects of higher incomes in relieving the worst aspects of poverty
- The capacity of dairying to create jobs
- The capacity of dairying to benefit the poor at low cost
- The importance of commercial approach to development
- The capacity of single-commodity projects to have multi-dimensional impacts
- The importance of getting government out of commercial enterprises
- The importance of market failure in agriculture
- The power & problems of participatory organisations
- The importance of policy.

Achievements of the "Amul Movement"

1. The phenomenal growth of milk production in India – from 20 million MT to 100 million MT in a span of just 40 years-has been made possible only because of the dairy cooperative movement. This has propelled India to emerge as the largest milk producing country in the World today.
2. The dairy cooperative movement has also encouraged Indian dairy farmers to keep more animals, which has resulted in the 500 million cattle & buffalo population in the country-the largest in the World.

3. The dairy cooperative movement has garnered a large base of milk producers, with their membership today boasting of more than 13 million member families.
4. The dairy cooperative movement has spread across the length and breadth of the country, covering more than 125,000 villages of 180 Districts in 22 States.
5. The dairy cooperatives have been able to maintain democratic structure at least at the grass-root level with the management committee of the village level unit elected from among the members in majority of the villages.
6. The dairy cooperatives have also been instrumental in bridging the social divide of caste, creed, race, religion & language at the villages, by offering open and voluntary membership.
7. The dairy cooperatives have been successfully propagating the concepts of scientific animal husbandry & efficiency of operations, which has resulted in low cost of production & processing of milk.
8. The movement has been successful because of a well-developed procurement system & supportive federal structures at District & State levels.
9. Dairy Cooperatives have always been proactive in building large processing capacities, which has further propelled growth of milk production.
10. The dairy cooperatives are among those few institutions in India, which still cherish a strong Cooperative identity, values and purpose. They still boast of idealism & good will of members and employees.
11. The dairy cooperatives have removed the poor farmers of India from the shackles of agents & middlemen and provided an assured market for their produce. As these are the institutions run by farmers themselves, it has also resulted in fair returns to the members for their produce
12. Dairy cooperatives have been able to create a market perception of honesty & transparency with their clean management.

Achievements of GCMMF:

- 2.8 million milk producer member families
- 13,759 village societies
- 13 District Unions
- 8.5 million liters of milk procured per day
- Rs. 150 million disbursed in cash daily
- GCMMF is the largest cooperative business of small producers with an annual turnover of Rs. 53 billion
- The Govt. of India has honoured Amul with the "Best of all categories Rajiv Gandhi National Quality Award".

- Largest milk handling capacity in Asia
- Largest Cold Chain Network
- 48 Sales offices, 3000 Wholesale Distributors, 5 lakh retail outlets
- Export to 37 countries worth Rs. 150 crores
- Winner of APEDA award for nine consecutive years.

Amul Brand Building

GCMMF (AMUL) has the largest distribution network for any FMCG company. It has nearly 50 sales offices spread all over the country, more than 3,000 wholesale dealers and more than 5,00,000 retailers.

AMUL is also the largest exporter of dairy products in the country. AMUL is available today in over 40 countries of the world. AMUL is exporting a wide variety of products which include Whole and Skimmed Milk Powder, Cottage Cheese (Paneer), UHT Milk, Clarified Butter (Ghee) and Indigenous Sweets. The major markets are USA, West Indies, and countries in Africa, the Gulf Region, and [SAARC] SAARCneighbours, Singapore, The Philippines, Thailand, Japan and China.

In September 2007, Amul emerged as the leading Indian brand according to a survey by Synovate to find out Asia's top 1000 Brands.

Products

Amul's product range includes milk powders, milk, butter, ghee, cheese, curd, chocolate, ice cream, cream, shrikhand, paneer, gulab jamuns, flavoured milk, basundi, Nutramul brand and others. In January 2006, Amul plans to launch India's first sports drink Stamina, which will be competing with Coca Cola's Powerade and PepsiCo's Gatorade.

In August 2007, Amul introduced Kool Koko, a chocolate milk brand extending its product offering in the milk products segment. Other Amul brands are Amul Kool, a low calorie thirst quenching drink; Masti Butter Milk; Kool Cafe, ready to drink coffee and India's first sports drink Stamina.

Amul's sugar-free Pro-Biotic Ice-cream won The International Dairy Federation Marketing Award for 2007.

Mascot

Since 1967 Amul products' mascot has been the very recognisable "Amul baby" (a chubby butter girl usually dressed in polka dotted dress) showing up on hoardings and product wrappers with the equally recognisable tagline Utterly Butterly Delicious Amul. The mascot was first used for Amul butter. But in recent years in a second wave of ad campaign for Amul products, she has also been used for other product like ghee and milk.

Advertising

Its advertising has also started using tongue-in-cheek sketches starring the Amul baby commenting jovially on the latest news or current events. The pun in her words has been popular. Amul outdoor advertising uses billboards, with a humorous take on current events and is updated frequently. The Amul ads are one of the longest running ads based on a theme, now vying for the Guinness records for being the longest running ad campaign ever with Smokey Bear. Sylvester da Cunha was the managing director of the advertising agency, ASP, that created, in 1967, the campaign.

Rivals

The success of Amul resulted in similar organizations being setup by state governments throughout India, most of which had reasonable success. Examples are Milma in Kerala, Vijaya in Andhra Pradesh, Aavin in Tamil Nadu, K.M.F (Nandini) in Karnataka, Sudha in Bihar, Omfed in Orissa, saras in Rajasthan, Parag in Uttar Pradesh, Verka in Punjab, Aanchal in uttaranchal, Vita in Haryana and others.

Other co-operative rivals of Amul include National Dairy Development Board (NDDB) (with its Mother Dairy and Sugam brands). With Amul entering the sports drink market, its rivals now include Coca Cola and PepsiCo.

In Popular Culture

The establishment of Amul is also known as White Revolution. The White Revolution of India inspired the notable Indian film-maker Shyam Benegal to base his film Manthan (1976) on it. The film starred Smita Patil, Girish Karnad, Naseeruddin Shah and Amrish Puri. The film itself was financed by over five lakh rural farmers in Gujarat who contributed Rs 2 each to the film's budget. Upon its release, these same farmers went in truckloads to watch 'their' film, making it a commercial success., the film was chosen for the 1977 National Film Award for Best Feature Film in Hindi. The Amul success story is taken up as a case study in marketing in many premier management institutes across the world.

The White Revolution ushered an era of plenty from a measly amount of milk production and distribution. Aside from the great measurable success that this project was, it also demonstrated the power of "collective might". A small set of poor farmers of Kheda district in Gujarat had the vision and foresight to act in a way that was good for the society and not for the self alone.

Bibliography

Ahlgren, G.H. : *Forage Crops,* New York, NY: McGraw-Hill, 1956.

Bogdan, A.V. : *Tropical Pasture and Fodder Legumes,* London: Longmans, 1977.

Clutton Brock Juliet : *Horse power: a history of the horse and donkey in human societies,* National history Museum publications, London 1992.

Clutton Brock Juliet : *The walking larder: Patterns of domestication, pastoralism and predation,* Unwin Hyman, London 1988.

Defoe, Daniel : *A Tour through England and Wales,* London: Everyman's Library, 1928.

Diamond, Jared : *Guns, Germs, and Steel: The Fates of Human Societies,* New York and London: Norton, 1997.

Fairey, D.T., & Hampton, J.G. : *Forage Seed Production of Temperate Species,* Farnham Royal: CAB. 1997.

Frame, J. : *Improved Grassland Management,* Ipswich: Farming Press, 1992.

Frame, J., Charlton, J.F.L., & Laidlaw, A.S. : *Temperate Forage Legumes,* Wallingford: CAB Interational, 1998.

G F Warren *: Farm Management,* Arise Pub, Delhi, 2008.

Gebauer, Anne Birgitte, and T. Douglas Price. : *Transitions to Agriculture in Prehistory.* Madison, Wis.: Prehistory Press, 1992.

Hacker, J.B. : *Nutritional Limits to Animal Production from Pasture,* Farnham Royal: CAB, 1981.

Hall S, Clutton Brock Juliet : *Two hundred years of British farm livestock,* Natural History Museum Publications, London 1988.

Hitchcock, A.S., & Chase, A. : *Manual of Grasses of the United States,* USDA, 1971.

Holmes, W. : *Grass, its production and utilisation,* Oxford & London: Blackwell, 1989.

J.M. Dhaka : *Economics of Agricultural Production and Farm Management*, Aavishkar Pub, Delhi, 2009.

Janick Jules, Noller Carl H., Rhykerd Charles L. : *The Cycles of Plant and Animal Nutrition*, Scientific American Books, San Francisco 1976.

Jones, M.B. & Lazenby, A. : *The Grass Crop,* London: Chapman & Hall, 1988.

Manger Louis N. : *A History of the Life Sciences*, New York, Basel 2002.

Raymond, F., Redman, P., & Waltham, R. : *Forage conservation and Feeding.* Ipswich: Farming Press, 1986.

Russell, Howard S. : *A Long, Deep Furrow: Three Centuries ofFarming in New England*, University Press of New England, 1976.

Skerman, P.J., & Riveros, F. : *Tropical Grasses.* Rome: FAO, 1989.

Skerman, P.J., Cameron, D.G., & Riveros, F. : *Tropical Forage Legumes,* Rome: FAO, 1988 .

Snaydon, R.W. : *Managed Grasslands,* Amsterdam & London: Elsevier, 1987.

Sweeney, Del. : *Agriculture in the Middle Ages: Technology,Practice, and Representation*, Philadelphia: University of Pennsylvania Press, 1995.

Whyte, R.O, Moir, T.R.G., & Cooper, J.P. : *Grasses in Agriculture,* Rome: FAO, 1959.

Whyte, R.O., Nillson-Leissner, G., & Trumble, H.C. : *Legumes in Agriculture,* Rome: FAO, 1953.

Index

□□□